SPEEDPRO SERIES

HOW TO BUILD, MODIFY & POWER TUNE

CYLINDER HEADS

PETER BURGESS

DAVID GOLLAN BEng. (Hon

VELOCE PUBLISHING PLC

PUBLISHERS OF FINE AUTOMOTIVE BOOKS

To our parents, families and friends, *Terminus A Quo*.

First published 1997, reprinted 1998, revised and updated edition 1999. by Veloce Publishing Plc., 33, Trinity Street, Dorchester DT1 1TT, England.
Fax: 01305 268864/e-mail: veloce@veloce.co.uk/website: http://www.veloce.co.uk
ISBN: 1-901295-45-1/UPC: 36847-00145-2

Readers with ideas for automotive books, or books on other transport or related hobby subjects, are invited to write to Veloce Publishing at the above address.
British Library Cataloguing in Publication Data -
A catalogue record for this book is available from the British Library.
Typesetting, design and page make-up all by Veloce on AppleMac.
Printed in the UK.

Contents

Veloce *SpeedPro* books -

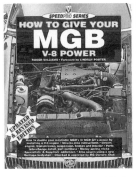

ISBN 1 901295 62 1

ISBN 1 874105 76 6

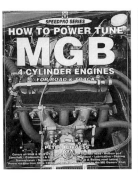

ISBN 1 874105 61 8

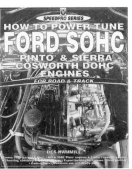

ISBN 1 874105 81 2

ISBN 1 874105 68 5

ISBN 1 901295 45 1

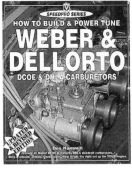

ISBN 1 901295 64 8

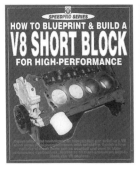

ISBN 1 874105 70 7

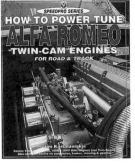

ISBN 1 874105 44 8

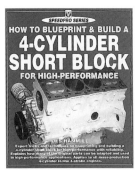

ISBN 1 874105 85 5

ISBN 1 874105 88 X

ISBN 1 901295 26 5

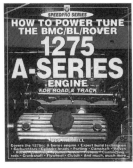

ISBN 1 901295 07 9

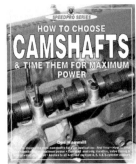

ISBN 1 901295 19 2

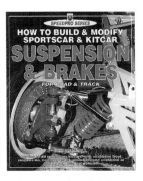

ISBN 1 901295 08 7

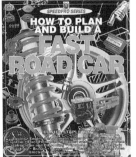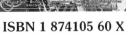

ISBN 1 874105 60 X

(visit www.veloce.co.uk
for further information
on all Veloce books)

- & there are more on the way

Introduction, Acknowledgements & About the Authors

INTRODUCTION

This book has been written to illustrate the theories and techniques which will allow an enthusiastic amateur to successfully modify cylinder heads. Applying the methods and techniques described will result in very noticeable increases in engine torque and horsepower.

Some of the theory may initially appear difficult to grasp, but patient reading and consideration will pay dividends - with practice you will be able to leave the 'professionals' standing.

Even if you don't wish to have a go at modifying your own heads, we hope this book will give a better understanding of the amount of work and skill that is involved when it comes to modifying cylinder heads successfully. At the very least, because you'll be aware of what to look for, you should be better equipped to go out and purchase a good quality, professionally-modified head that really will deliver the goods.

ACKNOWLEDGEMENTS & ABOUT THE AUTHORS

Peter Burgess built his own flowbench in 1985. A self-taught engineer, his background in psychology laid the groundwork for a scientific method approach and gave him the ability to harness his intuition in making engines work. He works holistically to modify cylinder heads.

Championship engines and heads built by Peter include - Rover (BMC/ BL) A Series, B Series, R, S and O Series, Rover V8, TR4, TR6, MG T Type, Pre war MG, Ford Pinto, Ford Kent Crossflow and pre-Crossflow, Toyota Starlet and Nissan Turbo, to name but a few. His favourite engine is the ex-Buick Rover V8.

Peter runs a small, friendly tuning business which enjoys worldwide recognition and distribution.

David Gollan is a qualified Mechanical Engineer. He and Peter met in 1990 whilst he was researching for his thesis on the effects of gasflow on bhp, fuel economy and emissions: they've been firm friends ever since. David keeps abreast of current automotive engine research by reading avidly all available SAE technical papers. He has a comprehensive knowledge of engineering principals which, combined with his natural flair for understanding airflow, meshes with Peter's skills and results in outstanding product performance. His favourite engine is the Ford OHC 'Pinto.' He and Peter spar on the racetrack on test days.

David works as a Hydrometrist for the Environment Agency and acts as Peter's consultant.

Our thanks to Liz Burgess for proofreading and stylistic elegance, to Keith Hippey for some of the line drawings and to Robert Day for developing and taking some of the photographs. Thanks also to Phil Gollan, Shaun Powell, Wayne Martin, Andy Fossey and the drag racers who gave us permission to take a few photographs.

Peter Burgess
David Gollan

Using This Book
& Essential information

USING THIS BOOK

Throughout this book the text assumes that you, or your contractor, will have a workshop manual specific to your engine for complete detail on dismantling, reassembly, adjustment procedure, clearances, torque figures, etc. This book's default is the standard manufacturer's specification for your model so, if a procedure is not described, a measurement not given, a torque figure ignored, you can assume that the standard manufacturer's procedure or specification for your engine should be used.

You'll find it helpful to read the whole book before you start work or give instructions to your contractor. This is because a modification or change in specification in one area will often cause the need for changes in other areas. Get the whole picture so that you can finalize specification and component requirements as far as is possible before any work begins.

Note that the term "thou" means thousandth of an inch and to "fettle" is to make small (usually) handwork adjustments until a component is a perfect fit.

ESSENTIAL INFORMATION

This book contains information on practical procedures; however, this information is intended only for those with the qualifications, experience, tools and facilities to carry out the work in safety and with appropriately high levels of skill. Whenever working on a car or component, remember that your personal safety must **ALWAYS** be your **FIRST** consideration. **The publisher, author, editors and retailer of this book cannot accept any responsibility for personal injury or mechanical damage which results from using this book, even if caused by errors or omissions in the information given. If this disclaimer is unacceptable to you, please return the pristine book to your retailer who will refund the purchase price.**

In the text of this book **"Warning!"** means that a procedure could cause personal injury and **"Caution!"** that there is danger of mechanical damage if appropriate care is not taken. However, be aware that we cannot foresee every possibility of danger in every circumstance. You should also appreciate that there is always a danger of cutting into a waterway during head modification and that such a mistake will usually mean that the head will have to be scrapped.

Please note that changing component specification by modification is likely to void warranties and also to absolve manufacturers from any responsibility in the event of component failure and the consequences of such failure.

Increasing the engine's power will place additional stress on engine components and on the car's complete driveline: this may reduce service life and increase the frequency of breakdown. An increase in engine power, and therefore the vehicle's performance, will mean that your vehicle's braking and suspension systems will need to be kept in perfect condition and uprated as appropriate. It is also usually necessary to inform the vehicle's insurers of any changes to the vehicle's specification.

The importance of cleaning a component thoroughly before working on it cannot be overstressed. Always keep your working area and tools as clean as possible. Whatever specialist cleaning fluid or other chemicals you use, be sure to follow - completely - manufacturer's instructions and if you are using petrol (gasoline) or paraffin (kerosene) to clean parts, take every precaution necessary to protect your body and to avoid all risk of fire.

Chapter 1
What is Horsepower?

For those of you already familiar with internal combustion engines and the four stroke cycle, please bear with us. Going back to first principals is the easiest means of linking together the various concepts and terms in a logical manner.

First, a brief description of the theory and application of the four stroke cycle; or Otto cycle, so called after the practical originator of the four stroke engine. By far the majority of reciprocating engines use this four stroke cycle, each cylinder taking four piston strokes - two revolutions of the crankshaft - to generate one power stroke.

The four cycles are -

1. Induction (or intake). The inlet valve is opened and the movement of the piston from Top Dead Centre (TDC) to Bottom Dead Centre (BDC) creates a vacuum that draws the supplied air and fuel mixture into the cylinder. To increase the amount of mixture drawn in, the inlet valve is normally opened shortly before the stroke begins and closed shortly after it ends.

2. Compression. Both valves are closed and the piston moving back up the cylinder reduces the volume between it and the cylinder head that the fresh mixture drawn in on the previous stroke has to occupy. This has the combined effect of compressing the mixture and further mixing the air and fuel together, which allows more of the energy available from the fuel to be utilised.

3. Power (ignition or expansion). The compressed fuel and air mixture is usually ignited before the piston reaches TDC (with both valves still closed) by a high voltage spark discharge from the spark plug. The resulting expansion of high temperature and high pressure gases produced by the burning air and fuel mixture then acts upon the top of the piston, which has now passed TDC.

The piston is pushed back down the cylinder, forcing the crankshaft to rotate. This is where the chemical energy in the fuel is converted into mechanical energy. The exhaust valve usually opens prior to the piston reaching BDC to begin the exhaust process and reduce the pressure in the cylinder.

4. Exhaust. The cylinder is cleared of the burned mixture remaining after the power stroke, initially blown out by the high residual pressure within the cylinder, with most of the remainder being swept out by the motion of the piston back up the cylinder as it approaches TDC. The inlet valve starts to open again as the piston nears TDC, the exhaust valve closes just after TDC and so the process begins again.

The more technical description used by the automotive engineering fraternity for the complexities of the four stroke cycle is: suck, squeeze, bang, blow.

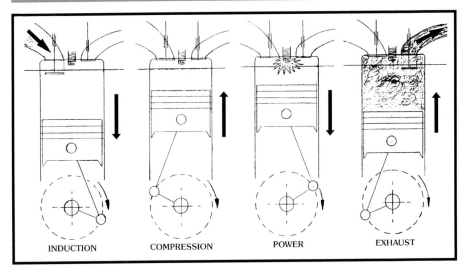

INDUCTION   COMPRESSION   POWER   EXHAUST

The four stroke cycle.

The fuel used by four stroke engines can be petrol (gasoline), methanol or ethanol (alcohol), nitromethane, diesel, fuel oil, liquid petroleum gas (LPG) or methane (natural gas).

The purpose of internal combustion engines is the conversion to mechanical power of the chemical energy contained in the fuel. In the case of spark ignition engines, this takes place inside the engine as the controlled burning of a mixture of air and fuel that is then converted into mechanical power by the engine's internal components. The engine's ability to undertake this conversion is known as its performance, and performance can be defined by the terms torque and horsepower.

Torque is the measure of the turning effort produced on the crankshaft by the pressure from the burning gases acting on the piston during the power stroke. To enable this measurement to be taken, the engine is coupled to a dynamometer, which is a device capable of exerting a force to oppose this turning effort (it in effect acts like a powerful brake). The amount of opposing force is given as a

reading in pounds feet (lb.ft.), or the more modern term of Newton Metres (Nm). If the engine and the dynamometer are running at the same rotational speed (revolutions per minute or rpm), then the engine torque must be equal to the dyno's opposing torque. Readings are taken over a range of different engine speeds and a torque output curve for that particular engine can be plotted.

The term brake horsepower (bhp - more modern definitions are PS or KW) is the power output of the engine at the flywheel (or output shaft), and is mathematically derived from the torque and rpm readings taken off the dyno.

The work output as measured by a dynamometer also allows the calculation of brake mean effective pressure (bmep). For normally aspirated engines, bmep shows the product of the ability to draw in an air/fuel mixture, how effectively the air is used in combustion, and how completely and efficiently the fuel is converted into energy. In short it denotes the average pressure acting upon the top of the piston over the complete four cycles of one cylinder. It

is a parameter which also includes the mechanical losses that occur within the engine due to friction. As the results, unlike the values for torque and power, are independent of engine size and configuration, they are useful for direct comparison with already established values from different engine types and designs.

Torque and bmep are directly related, so values plotted for both on graphs give curves having exactly the same shape.

The ability of the engine to utilise the heat released from the fuel is termed Thermal Efficiency. Unfortunately, internal combustion engines are not very thermally efficient. Only some 25 to 30% of the heat generated by burning the mixture is utilised in the form of useful work. The remainder is lost to atmosphere, mostly out of the exhaust pipe as a hot gas, the remainder via the coolant and oil or through heat radiation from the engine block.

A part of the power produced by the engine per cycle is used to draw in the fresh air/fuel charge on the intake stroke, compress it, and pump the burnt remains out on the exhaust stroke. Power is also utilised in overcoming the sliding and rotating friction of the internal mechanical components, such as pistons, rings and bearings, as well as to drive the engine accessories such as camshaft, distributor and oil pump. These losses are all grouped together under the heading of friction power. This friction power dissipates useful work as heat into the oil and coolant.

Overcoming drag from the oil (whipped up from the sump) acting on the crank and conrods creates another power sapping loss. It, too, is grouped under the heading of friction, but is also more commonly called windage loss.

Volumetric efficiency is the most fundamental and important parameter used for defining performance characteristics, as it dictates how much power an engine is capable of producing. It is a measure of the effectiveness of the engine's induction process, its ability to draw the air and fuel mixture into the cylinder. The incoming air has to make its way to the cylinder via the air filter, carburettor, throttle butterfly, inlet manifold - or the throttle body, plenum and runners in the case of fuel injection - the inlet port and finally past the inlet valve, negotiating bends and obstacles that can amount to a fairly tortuous route. All this can add up to quite a restriction to the amount of air that the engine can draw in. Typical values for modern engines range from 80 to over 90 percent volumetric efficiency, older designs usually manage around 60 to 70 percent efficiency. This means there is plenty of room for improvement!

Careful assembly and attention to detail can minimise the frictional losses mentioned previously, as well as creating an increase in bmep, which, in turn, will generate more torque and therefore horsepower.

As there is a direct relationship between the mass flow rate of air into an engine and the power capable of being developed, it is vital to get as much air as possible - together with the correct proportion of fuel - into the engine's cylinders. Or, to put it another way, improving the engine's volumetric efficiency is the key to good useable power gains, as the mass of air in the cylinder governs how much fuel

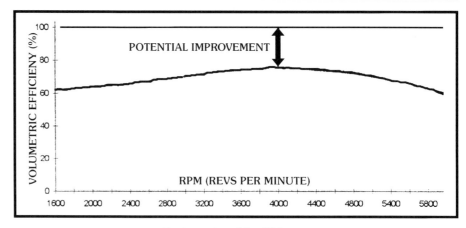

Engine volumetric efficiency.

can be burnt. Anything that makes life easier for the air to get into the cylinders results in improved volumetric efficiency, generating higher bmep, and so more torque and power - always provided other power losses have not escalated for some strange reason.

In the case of most older designs of engine, a well modified cylinder head offers the greatest all round improvement - to both volumetric and thermal efficiency - where the ports and valves have been improved or reshaped as necessary to smooth the passage of the air into the cylinder. Changing other components, such as the camshaft (which can be thought of as the engine's 'brain' in that it controls the inlet and exhaust processes) to one that opens the valves further and for longer, can also achieve reasonable power gains. But these gains generally occur at a higher rpm than for the standard engine which, in most cases, means sacrificing power at lower rpm.

The entire engine system really

needs to be considered as a whole, otherwise the performance gains possible from component changes may not be fully realised, their being limited by another area starting to cause a flow restriction or bottleneck. The 'Holistic' approach that the authors adopt will become more readily apparent in the following chapters.

Hopefully, you've now been given a reasonable overview of the various factors that are used to evaluate an engine's performance, and the effectiveness of any changes or modifications made. There are, of course, many more ways of mathematically modelling and evaluating an engine's internal and external processes, most involving complex formulae that are rather beyond the scope of this book. You may have noticed we've relegated our maths section to the back of this book - all that theory is fine in small doses, but what we're really interested in are the more practical methods of making horsepower!

Chapter 2
Types of Cylinder Head

The cylinder head can be made of cast iron or aluminium alloy and seals the top of the cylinders. Heads must be rigid enough to withstand the gas pressures exerted upon them during the engine cycle, constructed to contain passages for coolant flow (unless aircooled) and accomodate various other components - spark plugs, valves, etc., that are part of the ignition system and valvetrain. The majority of modern engines are of overhead valve design, which allows compact efficient combustion chambers and improved volumetric efficiency through better breathing.

Heads can be of overhead cam design, either single (SOHC) or double (DOHC) and these cams may operate the valves either directly, through action on bucket followers, or indirectly via finger followers or rockers. The alternative is to have the valvetrain operated via pushrods by a camshaft mounted within the cylinder block.

There has been a wide variety of combustion chambers designed and developed over the years. They have come about through the search for better breathing, improved combustion and more efficient, lighter valvetrains - reduced inertia allowing higher engine operating speed.

The various types of cylinder head are:

Inline OHV - Valves are generally arranged vertically, with bathtub, heart or kidney shaped combustion chambers.

Heron head - OHV engine with inline (vertical) valves and the combustion chamber in the piston.

Wedge - The valves are inline but angled from the vertical creating a wedge-shaped combustion chamber.

Pent-roof - The valve heads are opposed (stems in a V-formation) in a

Bathtub chamber.

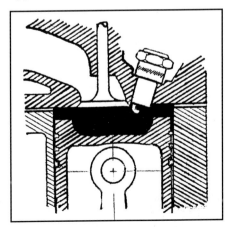

Chamber in piston.

Wedge head.

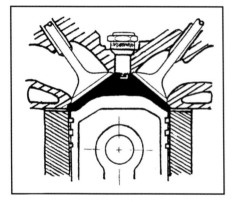

Hemi head.

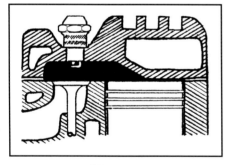

Sidevalve.

sloping sided combustion chamber. Increasingly used for modern four valve per cylinder layouts.

Hemi - The valve heads are opposed, as above, but the chamber is hemispherical.

L-type or flathead - An older

Pent- roof four valve per cylinder chamber.

design used for engines with the valves contained within the cylinder block (sidevalves).

SQUISH

'Squish' is the descriptive name given to the random movement of the yet unburned air and fuel mixture in the cylinder, caused by the piston nearing the cylinder head towards the end of the compression stroke. In most cases, a part of the piston crown and the cylinder head get pretty close to one another at Top Dead Centre (TDC), approaching at fairly high speed. Mixture caught between these rapidly closing surfaces is 'squished' or squirted out of the way. The easiest avenue of escape is into the main body of the mixture already contained within the combustion chamber, resulting in a useful increase in the movement of this fresh charge. The majority of the mixture is now confined to the combustion chamber in preparation for the ignition spark. This random, turbulent movement generated in the chamber helps the air and fuel further mix together, resulting in more complete combustion.

SWIRL (& TUMBLE)

'Swirl' is the name for the organised

movement of the air/fuel mixture in the cylinder. It is becoming an increasingly common feature of cylinder heads on modern engines as manufacturers strive to maintain or improve power outputs, whilst reducing fuel consumption and emissions. As with squish, swirl is used as an additional technique to improve the mixing of air and fuel, and help promote more complete combustion.

Swirl can be created by using the port design to direct the incoming mixture. The port can be either curved smoothly to direct the flow past the valve, or have a hump or deflector in it to force the flow out one way at the last minute, so to speak. The incoming

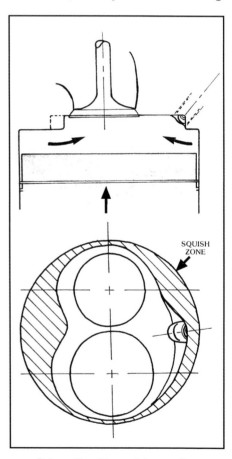

Schematic of how piston motion generates squish in a typical 'bathtub' chamber.

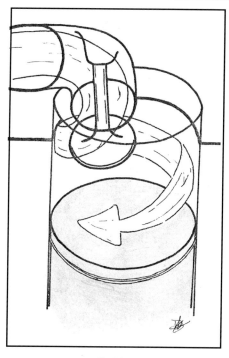

Swirl.

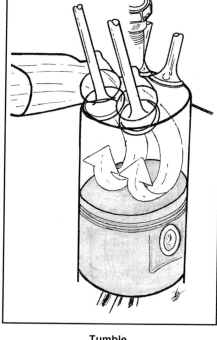

Tumble.

called helical ports, and create a swirl not unlike water going down a plug hole. While they have greater volumetric efficiency than the other types, as the flow is from around the whole of the valve, they are (fortunately) currently uncommon - they may prove difficult to modify!

Of course, what's left of the organised movement is disrupted as the piston travels back up the cylinder during compression, leaving the air/fuel mixing to random turbulence and any squish that may be generated around TDC.

While the idea of inducing greater motion to the incoming fresh air and fuel charge, with its resultant improved mixing and combustion, is a sound one, the slight downside is the reduced flow that it can cause. As the flow is usually aimed or forced to enter the cylinder in a certain direction, only a part of the valve's available flow area is fully utilised; the result is generally reduced volumetric efficiency.

You will need to consider carefully any modifications to swirl or tumble generating heads if you wish to maintain their characteristics. It is worth bearing in mind that hacking away at a head regardless may result in a decrease in combustion efficiency if the resulting air/fuel mixture is not presented in a manner the combustion chamber can cope with.

fresh charge is then given a bias by being made to flow predominantly past one side of the inlet valve, and so enter the cylinder at a tangent. It is then deflected by either the sides of the combustion chamber or the cylinder wall, and so spirals downwards around the cylinder to create a swirling motion. The intention is for this organised corkscrew rotation to be maintained as it travels down the cylinder.

Tumble is another form of swirl,

again generated by influencing the flow into the cylinder using the port's shape and direction. It differs from swirl in that the rotary motion of the mixture is parallel to the top of the piston, best described as the air and fuel being made to perform a horizontal barrel roll in the cylinder.

A further, more complex method of generating swirl is to use the port to create movement within itself, making the mixture rotate around the valve before it enters the cylinder. These are

Chapter 3
Theory and Thoughts

AIRFLOW REQUIREMENTS

A successful modifier of cylinder heads must combine the hands and eyes of a sculptor with the mind, curiosity and logic of an engineer and the tenacity of a terrier with the patience of Job. Quite an eclectic mix, but once you've had a go at head modifying you will possibly gain a better understanding of what it takes to modify heads for a living! You also need to be able to think from the point of view of the airflow, and not necessarily from a purely mechanical standpoint. Just because something looks really good and took a lot of work to modify, doesn't mean the airflow is going to concur!

Furthermore, in order to discover what does, or does not, work from the point of view of airflow, all the parts that are involved in getting air into and out of the engine should be tested on a flowbench. This is a piece of equipment that allows the quantity of airflow through a component to be accurately measured, and so any changes made can be individually evaluated to see if they work or not. Flowbench design is described in detail elsewhere in this book.

Refinement of the flow passages and component shapes involves establishing a baseline by testing the standard item in unmodified form. In the case of the cylinder head the port flow is measured at regular increments of valve lift. We usually use every 0.050in/1.27mm from zero to 0.5in/ 12.7mm (or more, if the valve lift of a particular engine is greater). This is followed by a series of gradual modifications to the head, each evaluated for its effects on flow, and hence used as a guide towards further modifications. It can often be a long, drawn out process, involving a great deal of painstaking work, to achieve worthwhile results. Consistently reproducing the end results of all this testing across all the ports in a head, and for many heads, needs a great deal of skill and experience (not to mention the benefits of the professional equipment we use). This is where the professional cylinder head modifier scores over the home tuner, and partly explains why modified heads cost what they do!

SUCK IT & SEE

Heads can be modified quite successfully without the use of a flowbench, but, in seeking the 'ultimate' flow from a head, who's to say you've gone far enough with the modifications or not? In most cases, it is necessary to go that one step further when developing a port shape, especially so in the case of race heads; when flow is lost, you know the previous step was the best. Often that one extra step finds fresh air in the form of a waterway or some other passage so the head must be scrapped, or kept as a pattern to remind one how far not to go next time. Obviously, it's not advisable to do this to a customer's cylinder head!

We have usually found during flowbench testing and evaluation of changes and modifications that, beyond a certain point, you start to lose flow at low valve lifts whilst continuing to gain at higher lifts. This trade-off of low lift to high lift flow must be considered, together with the ultimate use the completed head is to be put to. It is no use having massive flow gains at, say, 500 thou'/12.7mm lift if the cam and valvetrain only lift the valve to 350 thou'/8.89mm. Similarly, a huge port may flow well, but if the resulting gas speed is too low for proper cylinder filling, the engine will not give its best. In many cases a compromise is called for. Ideally, what we aim for is the greatest area under the curve (flow *vs* valve lift) compatible with the cam/valvetrain design and the engine's intended use.

Fortunately, you can bypass the research work and succeed in improving airflow, with a little effort and patience, by following our guidelines for straightforward and effective cylinder head modifications.

AIRFLOW FUNDAMENTALS

To make the reasoning behind our work a little clearer, we first need to cover a few fundamentals of airflow - which, in fact, apply throughout the entire inlet and exhaust system. These may appear somewhat odd at first, but will help in understanding our approach to modifying heads.

Airflow is more sensitive to shape than size, so big ports are not necessarily better than small ports at flowing air. Airflow also hates experiencing sudden changes in direction, volume and/or shape. These

concepts are supported by the fact that the areas of the port that are easy to get at normally have a small to moderate effect on airflow; it is more often those bits that are really difficult to get at that usually have the greatest influence on the head's airflow capability. Being able to modify those difficult bits in an effective manner is where the 'art' of cylinder head modification comes in.

As we have already said, improving the engine's volumetric efficiency is the key to good useable power gains. The words to note here are "useable power gains."

Volumetric efficiency is usually defined as the ratio of the volume of air drawn into the engine to the engine's swept volume. Filling such a large void (the cylinder) through a small passage (the port) very quickly means the incoming air is made to travel fairly quickly to do so. The average speed at which air travels through the port (its mean gas velocity) depends upon the volume of the cylinder, the speed (rpm) of the engine and the size of the port.

A small port feeding a large cylinder will have a high gas speed at a low rpm, it will be unable to supply sufficient air at higher rpm. Conversely, a very large port feeding the same cylinder will only achieve high gas speeds at high rpm and will have a very low gas speed at low rpm.

When you add fuel supply to the equation, things change again. The droplets of fuel supplied by a carburettor (or injector) need a reasonable velocity in order to keep them suspended in the airstream. If the velocity is too low, gravity takes over and they fall to the port floor forming puddles, which are difficult to get back into suspension and difficult to burn if it dribbles as far as the chamber (high hydrocarbons and

soot). Too high a speed and the droplets can become separated from the airstream when it has to negotiate a bend or obstacle, so the homogenous mix is lost and combustion problems arise.

The standard port dimensions are the result of the designers aiming to achieve the best compromise in terms of cylinder filling across the wide range of engine operating speeds.

At a certain engine rpm, and hence mean gas speed, the cylinder filling reaches its optimum, and peak torque is generated. Further increases in rpm produce higher gas speed in the port, up until a point where the flow becomes choked - it is just not possible for any more to get through. Around this point the engine produces peak horsepower. As the rpm continues to rise the cylinder demands for air can no longer be met, and so volumetric efficiency falls rapidly and, likewise, power decreases dramatically.

So, you do not want to go opening out the port without thinking about the engine's ultimate use. Big ports are fine for race engines where high rpm is called for and low speed driveability and emissions are not a consideration. Torque and some low speed capability are essential for an engine seeing use in an everyday road going vehicle.

It has been found that there is a correlation between the inlet mean gas velocity and engine volumetric efficiency. Our rule of thumb for playing the numbers game with port sizing is to use port velocities of around 80m/s (260ft/sec) for maximum horsepower and 50m/s (165ft/sec) for peak torque. These give ballpark figures to play with, though it has to be said that, at maximum power, modern multi-valve engines are probably achieving mean port velocities nearer 100m/s (300ft/sec).

The mean inlet gas velocity (V) can be calculated by -

$$V \text{ (m/s)} = (L \times N)/30000 \times (D/d)^2$$

Where -

    L = piston stroke (mm)
    N = engine speed (rpm)
    D = bore of cylinder (mm)
    d = port or valve throat diameter (mm)

(To convert from m/s to ft/sec multiply V by 3.28).

This example is for the Ford 2000cc SOHC 'Pinto' engine -
    L = 76.95mm
    N = 5500rpm
    D = 90.8mm
    d = 38mm

$$V = (76.95 \times 5500)/30000 \times (90.8/38)^2$$

$$V = 14.1 \times 5.71$$

$$V = 80.5 \text{m/s (or 264 ft/sec)}$$

As a reasonable guide, use this equation to calculate the gas speed at peak power for your engine using the factory specification (or use the factory specification of a high performance version of the same engine, if one was made). By rearranging the equation, you can calculate the port dimension to achieve the same gas speed (hence efficient cylinder filling) at a higher rpm or with an increase in bore size, or both, if you are intending the engine for fast road or race use.

Chapter 4
The Flowbench

What you cannot see is very difficult to measure. For instance, if you want to know the length of a rod you can measure it with a ruler; if you want to know the volume of a cylinder you can fill it with water from a measuring cylinder. If you wanted to know how much water would flow through a system you could fix a tank of water above it and time the water flow into a collecting bucket underneath. Then you could say "with a water supply z metres above the system the flow rate was x litres per minute" (or whatever units of measurement were being used). Unfortunately, we cannot put a bucket underneath a cylinder head and collect air! We have to make use of the fact that nature abhors a vacuum and that pressure differences will always try to equalise.

Building a flowbench is easy: calculating the airflow and understanding what is happening is harder.

The accompanying diagram shows the whole system in schematic form and can be split into its component parts as follows.

Air receiver
This is made from 1 metre/40in of 150mm/5.9in diameter plastic drainpipe (6mm/0.23in wall thickness). The bottom is sealed using a 6mm/ 0.23in thick plastic, or perspex, plate glued into place.

The Orifice plate/Orifice drum
Use a cylindrical drum of at least 300mm/11.8in diameter and at least 300mm/11.8in in length (eg. an oil drum). The drum has to be cut in half diametrically to accept the orifice plate.

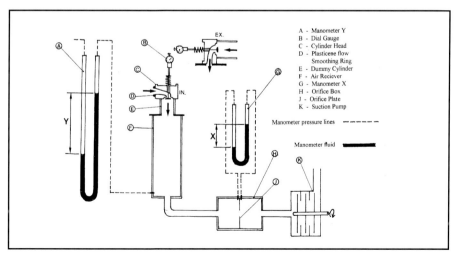

A - Manometer Y
B - Dial Gauge
C - Cylinder Head
D - Plasticene flow
    Smoothing Ring
E - Dummy Cylinder
F - Air Reciever
G - Manometer X
H - Orifice Box
J - Orifice Plate
K - Suction Pump

Manometer pressure lines   - - - - - -

Manometer fluid   ▬▬▬

Schematic of flowbench layout.

The orifice plate is cut from a sheet of 2.5mm/0.010in thick mild steel sufficient to overlap the diameter of the drum slightly. The centre of the plate is marked and then drilled/filed/ground to produce a hole 30-32mm/1.18-1.25in in diameter. The hole has to be as round as possible (it may be worth getting a machine shop to do this for you) and measured very accurately (to within 0.1mm/0.004in). This measurement is vital to the flow calculations, so keep a note of the exact dimension. It is also essential that the edges of the orifice are sharp and square and are **not** radiused.

The orifice plate is then welded to one half of the orifice drum, with the orifice as near central as possible and at right angles to the cylinder centre line.

Take care to ensure the weld is airtight and then weld the second half of the drum to the plate.

Flow bench pressure tappings

Use a short piece of thin metal or plastic tubing (maximum bore 6mm/0.236in) to make the tappings. Seal them so the inside ends are flush with the walls of the flowbench.

The tappings for manometer X are known as corner tappings - they must be placed each side as close to the orifice plate as possible.

The tapping for manometer Y should be placed fairly low down the air receiver. The other end of manometer Y is left open to atmosphere.

Suction pump flow settling drum

Use a drum similar in size to that used for the orifice box.

Manometers

To make the manometers use hard clear plastic tubing (eg. acrylic) of about 6mm/0.236in internal diameter.

For manometer X you will need two 1 metre/40in lengths of tubing, for manometer Y two 2 metre/80in lengths. The tubes should be mounted on wooden boards or wide planks for support. To attach them to the boards, mark the lines that the tubes will take (which must be parallel and about 130mm/5.2in apart), drill some small holes through the board and use thin gauge locking or copper wire looped around the tubes, passed through the holes and twisted together behind the board.

The interconnecting U-tube at the bottom can be made from small bore flexible plastic pipe, eg. windscreen (windshield) washer tubing, secured with a silicone sealant.

You will need four 1 metre/3.28ft rulers attached to the boards alongside the manometer tubes for the pressure drop measurements. A tip is to cut narrow slots in them and attach them to the boards with small screws, which can be loosened - the slots are to allow adjustment of the rulers' position. It means the manometer levels can easily be zeroed before flow testing begins.

Manometer pipes and damping

Flexible plastic tubing is needed to interconnect the orifice drum and air receiver to the manometers. Again, something like windscreen washer tubing is ideal as it is usually available in sufficient lengths. The connections must be airtight as any leakages will affect the readings. Silicone sealant performs the task admirably.

At a point some 600mm/23.6in from each manometer connection you will need to install pressure dampers. These are made from 0.75mm/0.030in capillary tubing (or similar) and need to be 100 times the bore diameter in length (75mm/2.95in) for consistent damping. These damp out the fluctuations in the pressure seen by the manometers, reducing the movement of the fluid, and so make taking readings easier.

Manometer fluid

Use water containing a few drops of washing up (dish washing) liquid, to prevent the water clinging to the walls of the manometer limbs. If you wish you can add a few drops of soluble ink to facilitate manometer readings (although we have found this to stain the inside of the tubes, making life awkward), though a white background for the manometer tubes also works well.

Suction pump

Use three houshold type vacuum cleaners (preferably drum type) and connect them to the suction pump flow settling drum. Alternatively, use one large industrial vacuum cleaner then the settling drum is not required. The total power requirement is around 2200 watts, or 3hp.

Connecting the flowbench

Use 75mm/2.95in rigid plastic drainpipe with the option of using the associated bends and angles or using flexible tubing for any curves. All the joints must be sealed and airtight - use adhesive, silicone or mastic.

Having described how to build and assemble all the component parts of the system we can move on to the next stage; crunching the numbers.

FLOW CALCULATION

What is air flow? Put simply it is the movement of air between two points. The movement is governed by two main factors; the square root of the pressure drop between the two points and the efficiency of the component being tested.

Quite a few people write articles on flow testing cylinder heads and they all seem to use different pressure drops to test the heads. This leads to controversy over which is right or wrong and confusion amongst those reading the articles as they have no direct means of comparison between results. Well, it doesn't matter one iota what pressure drop is used as one figure can be converted to another very simply. We happen to use a test pressure drop of 25 inches/635mm of water (H_2O) as the difference across the system being tested as this gives a nice graphical display of airflow.

The flowbench system described in this book does not work at any particular pressure drop; the maths converts the readings taken to the flow rate at 25in/635mm H_2O. As long as the manometer readings are above 50mm/1.96in then the flow rates will be fairly accurate.

As flow is directly proportional to the square root of the pressure drop, conversion from one flow rate in cubic feet per minute (cfm) to another can be worked out as follows.

To convert a flow value in cfm from pressure drop A to pressure drop B -

$$CFM\ B = CFM\ A \times \sqrt{(B/A)}$$

eg. 100cfm at 10in H2O to 25in H2O

$$cfm25 = 100 \times \sqrt{(25/10)}$$

$$cfm25 = 158.1$$

So 100cfm at 10in H_2O is exactly the same as 158.1cfm at 25in H_2O.

Simple calculation
Using the manometer readings and the

fact that the orifice has an efficiency of 62%, or has a coefficient of discharge (Cd) of 0.62, we can calculate the flow in cubic feet per minute using the following formulae -

Volume flow =
A x Cd x 213685.34 ....**1**
at 25in H_2O test pressure, 15 degrees centigrade and 1013.25 millibars atmospheric pressure.

Where -
A = area of valve or port in square metres.
Cd = coefficient of discharge of orifice or port.

The coefficient of discharge is a dimensionless number that is used to express flow efficiency, 100% being perfect.

To find the coefficient of discharge Cd -

$$Cd = \frac{Ao \times 0.62 \times \sqrt{(X/Y)}}{A} \quad ....2$$

Where -
Ao = area of orifice in m^2.
A = area of valve or port in m^2.
X = reading from manometer X in mm.
Y = reading from manometer Y in mm.

For example -
Using a 42mm valve, manometer X reading 500mm, 250mm reading from manometer Y and an orifice area of 0.0007 m^2.

Area of the valve = $(\pi D^2)/4$

Where -
D is the valve diameter in metres.
Area of the 42mm valve = 0.001385 m^2

Using equation 2 -
$$Cd = \frac{0.0007 \times 0.62 \times \sqrt{(500/250)}}{0.001385}$$

Cd = 0.443.

Or the valve is 44.3% efficient.

Using equation 1 above -
Flow = 0.001385 x 0.44 x 213685.34

Flow = 131.11 cfm @ 25in H_2O

If the valve was fitted to a port of 30mm diameter and the manometer readings were: X = 500 and Y = 250 with the orifice area 0.0007 m^2, what would be the coefficient of discharge for the port and what would be the flow rate be?

Area of port = 0.000707 m^2

Using equation 2 -
$$Cd = \frac{0.0007 \times 0.62 \times \sqrt{(500/250)}}{0.000707}$$

Cd = 0.868
Or the port is 86.8% efficient.

Now substituting Cd into equation 1-

Flow = 0.000707 x 0.868 x 213685.34

Flow = 131.15cfm @ 25in H_2O

The question was really a trick one as the manometer readings were the same, the calculated flow rate through the port and the valve had to be the same. Only the efficiency will change.

More detail
There is no point in going into a long winded explanation of the following formulae; suffice to say that they take into account atmospheric differences

on different days, and the fact that air is compressible.

By measuring how much water flows through the orifice box we can accurately determine the actual coefficient of discharge of the orifice.

We can calculate the theoretical discharge using the following formula -

Theoretical discharge =
$$_2\sqrt{(2 \times g \times h \times A)}$$

Where -
g = acceleration due to gravity 9.81m/s.
h = height of water above the orifice in metres.
A = area of orifice in square metres.

So with water supplied from a tank 2 metres above the orifice, and with an orifice 30mm in diameter -

Theoretical discharge =
$$_2\sqrt{(2 \times 9.81 \times 2 \times 0.0007)}$$

$$= 0.0044 m^3/s$$

If we measure how much water flows in one minute we can work out the actual discharge per second; for this example 0.163 m^3 would be collected in one minute.

This gives $\dfrac{0.163}{60}$ m^3 per second

$$= 0.00272 \ m^3/s$$

Coefficient of discharge =
$$\dfrac{\text{Actual discharge}}{\text{Theoretical Discharge}}$$

$$Co = \dfrac{0.00272}{0.00440}$$

$$Co = 0.62$$

Efficiency = 62%

The steps in calculating the flow rates are as follows -

(a)
$$M = A \times Co \times \Sigma \times {_2\sqrt{(2 \times ¶ \times \Delta p)}}$$

Where -
M = mass flow
A = area of orifice in m
Co = coefficient of discharge of orifice
Σ = compressibility factor (see e)
¶ = Density in kg / m^3 (see f)
Δp = change in pressure (N / m^3)
= 9.81 x X (reading from manometer)

This gives us the mass flow during the test. To convert to volume flow at standard temperature and pressure -

(b)
$$V° = \dfrac{M°}{1.225}$$

Where -
V° = volume flow in m^3/s at 15°C and 1013.25 millibars pressure (density 1.225kg/m^3)

(c)
To convert from cubic metres per second to cubic feet per minute -

$$cfm = V° \times 35.31467 \times 60$$

This will give us the volume flow at the pressure drop shown on manometer Y.

(d)
To convert test pressure to 25in. H_2O -

$$cfm25 = cfm \ test \times {_2\sqrt{(635/Y)}}$$

Where -
635 = 25 in. H_2O in mm
Y = reading from manometer in mm

(e)
Compressibility factor.

$$\Sigma = 1 - [(0.41 + (0.35 \times \beta\char`\^4)) \times (\Delta p/(Pa \times K))]$$

Where -
β = diameter ratio of orifice drum D to orifice d: or d/D.
Δp = pressure change in Newtons per metre squared N/m^2, given by reading from manometer X multiplied by gravity: X x 9.81.
Pa = atmospheric pressure in Newtons per metre squared N/m^2, given by -

millibars (from barometer) x 100 - (Y x 9.81) - (X x 9.81)

K = 1.404 which is the ratio of specific heat capacity at constant pressure to the specific heat capacity at constant volume.

(f)
Air changes density at different temperatures and pressures. This can be calculated as follows -

$$¶0.003482 \times \dfrac{P - (0.378 \times VP \times RH)}{(273 \times T)}$$

Where -
P = atmospheric pressure in millibars x 100 - (Y x 9.81).
VP = vapour pressure of water at test temperature T °C.
RH = relative humidity.
T = test temperature in °C.
The vapour pressure is found from tables, it represents 100 percent humidity. The actual humidity is found using a wet and dry thermometer hygrometer.
The atmospheric pressure is found using a barometer that reads actual pressure and not pressure at sea level.

As the figures used in the calculation are repeated in the various equations, and since the readings for X and Y change at each valve lift, a computer can be used to perform the calculations, either by writing a simple programme or by utilising the spreadsheet facility that is available with most computer packages.

USING THE FLOWBENCH

Leak free?

Before you can take any figures seriously the flowbench must be working properly; if there are any leaks then the figures will be too high, especially at small valve lifts. You can check by zeroing the manometers and then switching on the vacuum with the top of the air receiver sealed. Use a piece of wood or plastic, not your hand: the suction from the bench described will give a nasty blood bruise. A reading on manometer Y (the orifice drum) will mean a leak.

Dummy cylinder

Another equally important point is to match the tube that sits between the flowbench and the cylinder head to the exact size of the cylinder bore on which the head will be used. As the bore can shroud the valve, and so reduce the airflow, it is obvious that an oversized tube, or one that is offset (hence unshrouding the valve), will give flow figures that are too high. We cannot emphasise enough the importance of correctly simulating the real world situation that the head will encounter in use, only then will what you learn from the flowtests have any worth. You can use cylinder liners, plastic drain pipe, etc, turned to size.

A head gasket will help show where the head sits relative to the bore. A line can be scribed on the head face around the bore to help

Stock head under evaluation on flowbench.

with aligning the head on the flowbench dummy cylinder. We usually seal the dummy cylinder to the bench and to the head with plasticine.

Inlet port testing

The section of head to be tested needs to be assembled. For example, when testing the inlet of a standard head it would be assembled with the actual inlet and exhaust valves that are run as original equipment. The exhaust is helped seal to its seat with a thin smear of grease, and the standard valve spring is used to keep it shut.

The inlet valve will need a lighter than standard spring to allow it to be opened easily during the test, though not so light that the vacuum from the bench pulls the valve open any further than intended - yes, this device really sucks! The inner spring from a set of doubles is usually sufficient. Don't forget to install the spark plug - no laughing, it's easily overlooked - again a thin smear of grease on the threads will save having to wind it in too

tightly, just nip it up.

Exhaust port testing

This time the inlet is fitted with the smear of grease to help it seal, while the exhaust needs the lighter valvespring. For testing the exhaust port the head is mounted with the manifold face to the air receiver and the dummy cylinder bore is attached to the head face. This can be a bit more awkward to set up as the head is now sitting vertically rather than horizontally as before. The bench then pulls the air through the dummy bore, valve and port in the correct direction of flow.

You will also need to rig up some means of opening the valve. In many cases the standard rocker arms can be used, they can be modified with a long threaded bolt in place of the standard adjuster, the bolt providing an accurate means of achieving the desired valve lift for testing. Failing this, you will have to devise some means of lifting the valve to suit the particular application. Finally, a dial gauge and stand will be necessary to measure the lift. This must be fitted as rigidly as possible as any movement or flex will upset the accuracy of the tests. A small adapter on the end of the gauge may be necessary in order to reach the valve cap. We usually site the end of the gauge on the valve cap, close to the collets (valve keepers), so that errors arising from any movement of the valve are reduced.

Once everything is fitted to the bench and rigged up ready to go, set the dial gauge to zero and open and close the valve, to full lift and back, a couple of times by whatever means you are using. This is just to check that nothing catches, fouls or moves unduly and to allow the valvespring, cap and collets to settle in. You may find the dial gauge does not return to zero once or twice with the valve back on its seat, probably due to components settling. Once you are happy with the set-up and the dial gauge is at zero with the valve shut, zero the manometers.

We also fit a flow smoothing ring to the manifold face of the inlet port to help the airflow (see photo). It provides a constant for the flow tests and negates any sharp edge or ninety degree bend problems the air may experience at the port entry. This is just a rolled out length of plasticine shaped to suit the mouth of the port, which is a cinch to remove and refit for subsequent tests.

Switching the flowbench on with the valve closed will show up any leaks by giving a reading on manometer Y. These are best cured before testing commences.

Valve curtain area

Before proceeding further we now require a brief burst of theory, in order to help define test parameters.

The region around the valve through which the air has to pass is called the curtain area. This is the valve circumference times the valve lift. So the curtain area obviously increases with increasing valve lift. When the valve lift is equal to a quarter of the valve diameter (0.25 x D) the curtain area equals the valve area and the valve no longer represents the limiting factor (it should no longer present an obstruction) to flow.

Theoretically, a valve flowing perfectly would achieve 100% efficiency at this point. Since no more flow would be possible, any more valve lift would be pointless. However, 100% flow efficiency is practically unobtainable and so extra valve lift is used to extend flow capability.

It is useful to calculate the 0.25 x D value as it provides an arbitrary

minimum lift to open the valve to during the flow test. However, it is useful to continue testing beyond this point as it helps to develop a more complete picture of how the particular port/valve test combination behaves.

In most cases we find flow figures are best taken every 0.050in/1.27mm of lift up to, say, 0.5in/12.7mm lift, as for the majority of heads it is well above the 0.25D point. It is not often necessary to lift a valve more than this as flow increases beyond this point are generally very small. However, for some of the much larger valve sizes used for race heads (or some US V8s) you will need to lift the valve further.

It should be borne in mind that most cams/rockers do not lift the valve over 0.5in/12,7mm. For those engine combinations that do, use the 0.25D value or the physical valve lift, whichever is greater, as the minimum flow test valve lift - then go 0.2in/5mm beyond that, if possible.

Until the figures from the bench begin to make sense it is best to repeat the flow tests a couple of times and average the flow for each increment of valve lift. Check and question everything, even to the extent of removing the head from the bench and starting the testing afresh, should you be unsure of the validity of the figures. You'll learn a lot more by "how and why?" questioning and thinking things through.

THE FLOWBENCH EXPERIENCE

The good

A simple mistake can give rise to distorted figures - what seems like a good modification may even be losing flow. There is no substitute for experience.

For instance, flow development on a Triumph TR4 cylinder head

progressed steadily with the expected results, that is, after impressive initial gains, further modification followed the usual law of diminishing returns until the relatively simple modification of shortening the valve guide 0.125in/3mm increased flow by an astonishing 15%. Flow figures were rechecked and still showed the same gain. Through carefully checking everything, it turned out the head had been placed on the dummy bore incorrectly - equivalent to offsetting the head 0.12in/3mm from its usual bore position. This put the valves nearer to the centre of the cylinder and hence reduced the effect of bore shrouding. This "mistake" was deliberately duplicated on a race engine and the power increased from 127 to 150bhp at the wheels!

The moral is that you can, and must, learn from your mistakes. Usually you will learn what not to do, however, occasionally as above you will learn what you can do.

The bad

Whilst on the subject of mistakes it cannot be assumed that an increase in flow has to give rise to more power. All things being equal the power should increase in proportion to airflow, as long as the valve lift and duration as well as the induction and exhaust components can handle the increase. It must be borne in mind that, when you start altering the port and valve dimensions dramatically, you are moving further and further away from the designer's original parameters. You cannot put in a big valve and hog out the port and expect more power without flow developing the product and then power testing it.

As an example, we were working on a BMC MGB 1800 B-series four cylinder head, which has two individual inlet ports feeding a pair of valves - termed "siamesed ports." In trying to improve the chances of swapping airflow from cylinder to cylinder between siamesed pairs we decided that altering the wall that divides the two valves could reduce the time taken for air to change direction from one cylinder to the next.

The port divide was ground back to be similar to that used in the 1275 BMC A-series engine head, which also has siamesed inlet ports. Flow went down. It became apparent that flow was separating from the port walls at the point where the port expands to feed the valves. By carefully filling the port with epoxy resin filler, the abrupt expansion was reduced. Result - flow rates were the best we had ever achieved through a siamesed inlet port. However, when testing on the rolling road a power output of 140bhp at the wheels at 5000rpm (peak power was expected to be at 5800rpm with the component combination) reduced to 10bhp at 5100rpm! It seems that each valve opening and closing sets up a 'wave' that travels across the port to affect the opposite paired valve, forming something akin to an impenetrable barrier to airflow into the cylinder. This 'wave' doesn't occur when the port divide is in place - as originally designed by Weslake. The dimensions of the port divider between the valves is critical - an A-series head doesn't need the divider, an MGB does. Weslake probably discovered this when he designed the B-series engine, after he designed the A-series! Do not assume that the person who designed the head was stupid and knew less than you.

The ugly

Not exactly ugly, but the poor old Ford Crossflow head that suffered an intense flowbench development 'frenzy' was not an overly pretty sight once the dust had settled. With the Crossflow the ports are biased, so flow enters the cylinder heading into the cylinder wall (generating swirl?). This was deemed a problem as this should be bad for flow, an opinion various tuning magazine articles supported. Umpteen permutations were tested, altering the port to bias the flow away from the chamber wall, straightening the port shape - even to the extent of removing the port altogether and replacing it with a straight tube, something not wholly feasible for an engine (expensive, but not impossible!). During all this testing it was eventually decided to eliminate the effects of the cylinder wall shrouding simply by fitting the head to the dummy cylinder, with the inlet valve positioned smack bang in the middle of the bore - obvious really, but the simple checks are often overlooked. Flow was lost, returning when the valve was repositioned in its correct location. It turned out that there was no need to go to the trouble of reconfiguring the port shape, a rather complicated and time consuming process (read - expensive) to do it properly; conventional modifying methods worked well enough when applied to this head.

A lot of work just to prove you must take what you read or are told with a pinch of salt, and ideally find out for yourself. Don't overlook the simple checks either! Being versatile in your approach to testing means you can check many more scenarios than you may think.

Next, a quick example of flow development relating to seat width. Sometimes it is best to run a very wide (for a race head) valve seat. For instance, we developed a 47mm/ 1.85in inlet valve 3.5 litre Rover V8 racing engine with 1mm/0.039in 45° seat/60° throat cuts. Development reached a stage where at 800 thou'/ 20.3mm lift the flow was sufficient to give 385bhp (rule of thumb) with good flow up to that point. However, the cam to be used would only give 550 thou'/13.97mm valve lift, which worked out to a theoretical 305bhp with the flow available. A further problem was that, in trying to improve the short side of the port, it transpired that a waterway was too close for comfort (poking hard with a screwdriver we found fresh air!) making development work on the actual heads themselves difficult. Bearing in mind the two problems, we reworked another port to the same specification, but this time with a 1.5mm/0.059in 45° seat and a 2mm/ 0.078in wide 60° throat cut, leaving more meat on the short side. Flow above 600 thou/15.24mm, was rubbish compared to the original, but the bulk flow (given by the area under the curve of a flow *vs* lift graph) was enough for a theoretical 340bhp. The engine actually delivered 320bhp 6700 rpm) when dyno tested, with a slightly milder cam than we assumed would be used, and produced a very wide torque band (over 250lb.ft from 3500 to 6500, peak 283 at 5000rpm).

Chapter 5
Development Pointers

For a head you cannot afford to ruin, it's always a good idea to spend a little time poking, probing and generally investigating the design and layout before removing metal. By looking closely into the waterways and exploring, where accessible, with your fingers or a small screwdriver, you can get a feel for the contours and material thickness of the ports. This will help when it comes to metal removal as you should have some idea where any potential trouble spots are likely to be. It's important that the head is really clean as muck and corrosion make examination difficult.

Caution! - castings that appear outwardly similar to others from the same engine series, may have been subject to design modification and revision during the engine series' lifetime. Heads from different batches may have internal differences that create different sets of work constraints to those you have established with a similar head. If you assume they are identical and work as such, you may end up finding fresh air where once there was metal.

If you can sacrifice a head (in the interests of science, naturally) that is thought to be identical to the one you will modify, then drilling holes in and around the port will reveal the material thickness of the walls and around the throat area. Any errors made during testing and development contribute to the learning process, so a sacrificial head will allow you to see how far you can go with the modifications and you'll have several ports to try different ideas on.

PORT & VALVE GUIDE BOSS

The port must be smooth with no irregular lumps or dips (excrescences or inclusions) or sudden changes of size. Valve guide bosses impinge upon airflow so a good designer will take this into account and form a guide boss, in the shape of a tapered ramp, for almost the entire port length, so guiding the air gently around the obstacle. Usually the area of the port around the guide (where it protrudes into the port) will be bigger to compensate for the obstruction. If the boss is an isolated lump, airflow will probably improve if it is removed.

VALVE THROAT & SEAT

The throat must blend into the valve seat and port without any sharp turns. The seat needs to be of the three-angle variety with the top cut blending smoothly into the combustion chamber and the bottom cut radiusing into the throat. Some port configurations benefit from having very little or no bottom cut on the short side turn, others may prefer a wide cut all the way around. The permutations are fascinating.

VALVE SEAT WIDTH

There is no hard and fast rule regarding seat width and valve size. Obviously, a small valve would end up with a very narrow throat area if the seat was any bigger than, say, 1mm/0.039in. Generally, the thinking is that a wide seat gives good low to mid-lift flow, whilst a narrow seat gives better mid to high-lift results. However, in reality, whether this holds true or not depends on the original design of the head, particularly the port/throat configuration. Experimentation is the only way to find out.

You also need to consider the mechanical side of the seat's role. When the valve is closed heat is transferred through the seat interface from the valve to the head. If the valve gets too hot it may be damaged, or create another problem in the form of pre-ignition/detonation. A wider seat gives a greater contact area for conducting the heat away, but too wide a seat can reduce the flow performance. A compromise situation.

A seat can also become damaged or worn during its life. A very narrow seat will have reduced longevity compared to that of a wide seat. Grit or tiny carbon particles can become caught between the valve and the head, causing cratering and other damage which, with a narrow seat, would jeopardise its sealing performance. Not a problem for a race application where the head(s) only need to survive one season (or less) before overhaul. For road use, though, the chore of having to pull the head(s) off to freshen up the seats every few thousand miles would soon reduce to insignificance any performance gains from very narrow seats. Again, a compromise situation.

Valve to head heat transfer is an especially important consideration when modifying heads for turbocharged applications as the exhaust valve has an increased thermal load placed upon it. Here, it's best to err on the side of mechanical longevity.

We tend to favour a 1.3mm/0.050in seat width for most applications, giving a good flow/heat dissipation compromise.

Occasionally, if a throat is already wide, it's impossible to cut a very wide 45° seat and achieve a worthwhile 60° bottom cut, so a 1mm/0.039in 45° seat is cut (even for road use) as power and torque are much greater with a 60° bottom cut.

VALVE SHAPES

Valve shape will vary depending on the angle of approach of the port. A 'sidedraught' 90 degree bend-type port usually flows better with a fairly flat backed (7 to 10 degrees) valve. With a 'downdraught' port, where the flow impinges directly on the back of the valve, a tulip-shape will help to steer the air around the valve and out of the throat.

Plasticine (or modelling clay) is a good development tool for experimenting with alternate shapes for the backs of valves. Car body filler is another option, a harder wearing medium that's readily reshaped and can be made to merge with the surrounding material, giving a smooth transition with no ridges or steps.

CHAMBER SHROUDING

When possible, the inlet valve should be around a quarter of its diameter (0.25D) away from the combustion chamber wall. The exhaust should be around two tenths of its diameter (0.2 D) away from the combustion chamber wall.

If the valve throat is fairly vertical, flow benefits if the combustion chamber roof is flat until the chamber wall is reached. This allows the air to expand freely around the valve periphery.

If the throat is more angled, flow doesn't seem to increase if the roof is flat - a curve is as good.

As a bonus, near vertical chamber walls create more turbulence and hence improved combustion, giving more power whilst requiring less ignition advance.

Caution! - A word of warning with regard to modern fuels: too much unshrouding can cause detonation because the combustion flame has further to travel to consume the fuel (eg. standard Ford Pinto, Vauxhall Cavalier 1.8 eight-valve engines, and Peugeot 1.6/1.9 eight-valve engines.

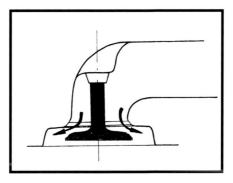

With a vertical throat, a flatter valve back and flat chamber roof allows good airflow past valve head.

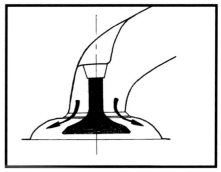

With a more angled throat, airflow benefits from tulip valve shape and curved chamber roof.

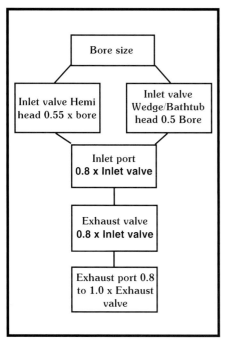

```
        Bore size

Inlet valve Hemi      Inlet valve
head 0.55 x bore      Wedge/Bathtub
                      head 0.5 Bore

        Inlet port
      0.8 x Inlet valve

       Exhaust valve
      0.8 x Inlet valve

     Exhaust port 0.8
     to 1.0 x Exhaust
           valve
```

A starting point. As a rule of thumb, these are good relationships between cylinder, valve and port sizes.

VALVE FLOW POTENTIAL

The problem with estimating the flow potential for a valve is that the flow only passes through the curtain area, the product of the valve circumference and lift. A rule of thumb we use is to assume a flow efficiency of 95% for the valve area. The derivation of this is based upon the flow potential per square millimetre of valve curtain area, a bit too long-winded to explain here, but the numbers plug nicely into the flowbench maths.

Most theoretical models for flow calculation use the valve throat area as the narrowest point through which the air must pass. We can easily calculate the valve seat inside diameter, assuming a 45 degree seat is being used, which is near enough the throat diameter for these purposes; the alternative is to measure.

Valve seat id. = valve dia. - (seat width x cos. 45)

eg. for a 42mm diameter valve with a 1.5mm seat -

Valve seat id. = 42 - (1.5 x 0.707) = 41.1mm

From this the curtain area can be calculated -

$$curtain\ area = \pi \times Diameter \times Lift$$

For the calculated 41.1mm at 1mm lift. For the maths the numbers must be in m^2.

$$Area = (41.1/1000) \times \pi\ (1/100\ )$$

$$= 0.000129 m^2$$

Using the flowbench maths -

$$cfm = A \times C \times 213685.34$$

and assuming 95% efficiency -

$$cfm = 0.000129 \times 0.95 \times 213685.34$$

$$= 26.2 cfm\ @\ 1mm\ valve\ lift$$

Curtain area -

$$0.25 \times 41.1 = 10.275mm\ lift$$

Valve flow potential -

$$cfm\ potential = valve\ flow \times valve\ lift$$

$$= 26.2 \times 10.275$$

$$= 269.32 cfm\ @\ 10.275mm$$

This may seem like excellent flow but, remember, it's only our 'rule of thumb' theoretical flow potential and there is still another factor to be calculated.

Port size & flow restriction

If you reach a point in the flow development where flow does not increase, even with a bigger valve, it could be that you have reached the maximum flow capability of the port. The diameter of the valve stem may also need to be taken into account as it reduces the available area for flow.

If you assume, say, 95% efficiency, you can calculate the flow potential of the port using the flowbench maths.

$$cfm = A \times 0.95 \times 213685.34$$

Where A is the area of the port in square metres (use the narrowest point , eg.-

32mm diameter port, area is $0.000804 m^2$

Therefore -

$$cfm = 0.000804 \times 0.95 \times 213685.34$$

$$= 163.3 cfm$$

Or, including a valve stem of 8mm diameter -

Port area - valve area = 0.000804 - 0.000050

$$= 0.000754 m^2$$

Therefore -

$$cfm = 0.000754 \times 0.95 \times 213685.34$$

$$= 153 cfm$$

In both of the foregoing examples there is much less flow capacity. A 42mm inlet valve would be port restricted and, no matter how much you modify the valve, throat and chamber, you will not increase flow! (See the flow development of the 351 Ford V8).

The same maths can be applied to inlet manifolds.

By rearranging the previous equation, a port dimension can be calculated for a given flow value.

For example, what increase in area would compensate for the losses caused by the presence of the valve stem?

Rearranging -

$$cfm = A \times 0.95 \times 213685.34$$

To give -

$$\frac{cfm}{0.95 \times 213685.34} = Area$$

Say the port with the valve (from previous example) flows 153cfm and you want 163.3cfm, you'll need the difference in flow on top of the flow required, ie, (163.3 - 153) + 163.3 = 173.6 required mathematically.

$$\frac{173.3}{0.95 \times 213685.34} = Area$$

$$= 0.000855m^2$$

Dia. in metres = $2\sqrt{((0.000855 \times 4)/\pi)}$
$$= 0.03299 \times 1000$$

$$= 32.99mm \ diameter$$

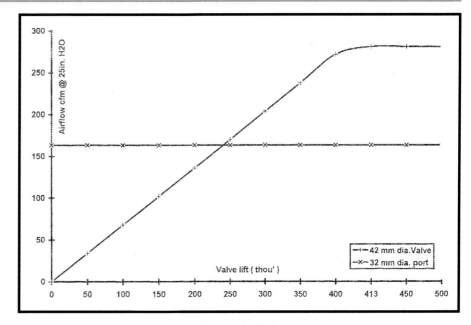

Flow potential.

The figures for the port and valve flow potential can be combined on a graph to give a visual presentation. By allowing comparison of the theoretical flow with actual test results, the figures can be used as the target/guide for your cylinder head flow development program.

Obviously the maths backs up the 'bigger is better' myth for inlet port sizing, but, remember, it does not consider the flow velocity which is an essential part of good cylinder filling. By chasing the valve flow numbers by opening up the port size, velocity will be lost.

Chapter 6
Head Work

INSPECTION AND CLEANING

Inspection

With the head(s) off the engine and stripped, you can perform the first visual inspection to see if all is well and the head is actually worth working on or, if necessary, find another. You can also carry out these checks if the head is only off the engine for a decoke, and so nip any potential problems in the bud. The studs are removed from the head and manifold faces using a stud extractor: a low-tech alternative is to lock two nuts together on the stud and unscrew with the lower nut. Any studs that snap off will have to be drilled out and have a thread insert fitted. Replacement thread inserts (Helicoil or similar) and their fitting tools are available from engineering suppliers and some tool stores. If you don't want to buy the fitting equipment, most engine reconditioners or jobbing shops will be able to fit thread inserts for a small fee.

Good quality valve spring compressors make disassembly and assembly much easier.

Helicoil installing tools and Helicoil inserts.

If a crack is suspected, get it checked with a dye penetrant (available from engineering trade suppliers), or try your local reconditioner. A common place for cracking to occur is in the combustion chamber. Check the narrow bridge of metal between the inlet and exhaust valve in **all** the chambers. You'll need to remove any carbon from this region to allow proper inspection; coarse emery paper

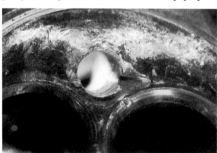
Here, a cambelt failure led to a coming together of piston and valve. The valve snapped, damaging the inlet seat and spark plug. Removal of the damaged plug stripped the thread, necessitating a Helecoil insert to be fitted.

After the Helicoil repair seen here, the valve seats were recut and the head reclaimed.

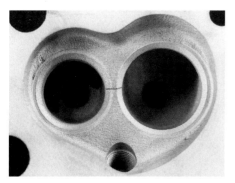

Crack between valve seats.

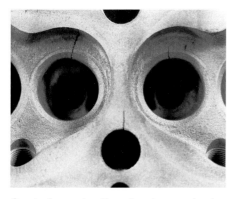

Cracks in combustion chambers and valve seats.

is ideal for this purpose. Lesser cracks can be repaired using replacement valve seat inserts, but you'll have to go to a head specialist or engine remanufacturer to get the work done.

Cracks can also go from the exhaust valve seats out across the combustion chamber, towards the chamber wall. These can be repaired

by fitting exhaust seat inserts, provided the crack is not too bad. If it has actually reached the wall of the combustion chamber, the head is **not** repairable and must be scrapped.

Check for cracks very carefully as there is no point expending a lot of time and effort on a head that only ends up being scrapped due to a severe crack.

Cleaning

Having passed the initial checks, the head needs to be thoroughly cleaned. For cast iron heads access to a chemical wash tank is useful. The hot caustic solution in a chemical tank, as well as removing caked-on oil, gaskets and any loose paint, has the big advantage of cleaning out from the waterways in the casting some of the corrosion and accumulated detritus.

Caution! - Aluminium heads need to be cleaned in a different chemical solution, such as Trichloroethane, as a caustic solution would very rapidly eat away at the metal, giving off an explosive gas in the process! This solvent acts in a similar manner to that used for cast

Head being removed after soak in caustic tank. After which ...

... it is hot washed to remove all remaining chemicals.

Once dry the head goes to the blasting cabinet. (We don't usually leave the door open!)

iron, dissolving away any oil sludge and helping loosen carbon in the ports and corrosion in the waterways.

Note that corrosion in the waterways acts like an insulating layer, reducing heat transfer to the coolant, and creating hotspots where coolant flow is restricted, which may bring about cracks. Removal of this corrosion has a very beneficial effect on cooling around the ports and chambers by allowing the water to circulate more freely.

A stiff brush and an old plastic bowl containing a cleaner such as Gunk works well at removing all the old oil and bits of gasket from the head if a dip in a caustic or solvent tank is not possible: it just requires some hard work. Poking about with a screwdriver helps dislodge any muck from those hard to reach places. The best do-it-yourself method of removing internal corrosion is to blank off the thermostat and heater take-off holes and fill the waterways with a corrosion remover and descaler that's designed to clean central heating systems. Leave it to soak for a good while and then thoroughly flush the head out.

When it's completely dry, the head needs grit blasting to remove all the remaining baked-on carbon from the ports and chambers. A wire brush

Half and half to highlight the muck remaining, even after chemical cleaning.

Check out the carbon in the ports and chambers (right); even with blasting facilities it takes time to get a head properly clean (left).

mounted in a drill will also dislodge much of the carbon if blasting is not available. Once clean, the head is ready for a further visual inspection, just to ensure nothing was missed during the initial checks.

SAFETY FIRST!

Before we get down to the nitty gritty of how to modify heads, a word about safety is vital. **Warning! - Always wear eye protection when grinding heads.** Safety glasses are the absolute minimum requirement, but even then the dust and grit whizzing about while grinding can find its way around them and into your eyes. Goggles or a full face visor are better. Wearing a dustmask (respirator) is also advisable, as inhaling the dust is not to be recommended. Ear defenders are optional (it depends on how noisy your equipment is) but are generally a good idea as the noises from the grinding can be rather irritating.

Overalls are also to be recommended, the swarf, grit and dust soon renders things filthy, finding its way down your collar and up your sleeves surprisingly easily. Anything to reduce the laundry is to be commended!

Strong footwear (safety shoes/ boots) should also be worn; you don't want to accidentally drop a cylinder head on unprotected toes, it may get damaged!

Leather gardening/riggers gloves will provide hand protection, though you'll have to take them off occasionally to feel your work.

GRINDING GENERAL

Having equipped yourself with safety gear, you need to find somewhere suitable to work, preferably well-lit.

Typical fabricated head stands. Rubber tubing (rear) protects head.

One good tip here - an old skeleton (Anglepoise) desk lamp makes an excellent movable light source for illuminating awkward-to-see parts of the head. Grinding dust makes a dreadful mess, and spreads over quite an area (especially when using air grinders), so bear this in mind when choosing a place to work. Be warned; the dust can and will sneak into the light fitting, shorting out the contacts and blowing the bulb or, worse, the fuse in the supply. **Warning!** - As a safety precaution take the bulb out before starting work and wipe or blow the grit away. Disconnect the lamp from the mains before cleaning the bulb holder; a belt from the mains will really put a crimp in your day!

Some form of support for the head whilst it's being worked on is advisable. It will make positioning for working so much easier than if you have to keep humping the head around the bench. Stands are available from some aftermarket head manufacturers, or you could fabricate a pair from some steel rod (see photo

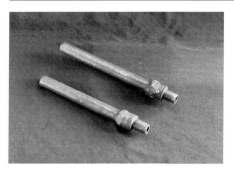

An alternative is to modify some spark plugs and then ...

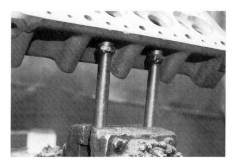

... the head can be held secure in a vice.

for an idea of shape). Cover the arms of the stand with lengths of rubber tube of an internal diameter to give a snug fit. The tube helps prevent damaging the head's surfaces as it's moved about, and provides a non-slip surface so heads won't move around so much while you're working on them. The stands will also come in handy for buretting and assembling the heads later on.

An alternative is to remove the ceramic and internals from a pair of spark plugs (the correct size to suit the head) and extend the metal bodies by welding on a couple of lengths of steel rod or tube (see photo). When the plugs are threaded into adjacent plug holes (the middle pair) the rods can be clamped in a bench vice. Make the rods long enough to allow for some height adjustment when in the vice, so the head can be positioned comfortably for working on.

GRINDERS

There are two ways to modify a cylinder head: the right way and the wrong way.

The wrong way

Trying to use an electric drill or a flexible shaft driven by an electric drill is a waste of time. The drill only operates at a few thousand rpm, insufficient for the stone or carbide cutters to work effectively. The design of the cutting surfaces of the tools means they require high rpm to function effectively. The authors rarely use less than 12,000rpm, which would suit a 1in/25.4mm ball carbide cutter. Anything smaller, for example, a 0.5in/ 12.7mm carbide, needs approximately 17,000rpm, while the smaller mounted points normally need to be used at around 25,000rpm.

Stock (metal) removal is very slow with an electric drill and the mounted points wear out at a dramatic rate at slow speeds. Also, an electric drill is not usually engineered to run flat out for very long: 5 to 10 minutes is usually enough to get most drills very hot! The bearings are not intended to take the side loads caused by grinding, and can wear out rapidly if used in too brutal a manner.

Furthermore, the chuck, be it on the drill or on the end of a flexible shaft, reduces the reach into the ports dramatically as well as limiting visibility compared to a collet type chuck.

The right way

Having decided to equip yourself properly with a die grinder of 20000rpm potential (a die grinder is the tool used to hold the burrs or mounted points - used in industry for grinding and finishing dies and so ideal for porting heads), you will have to decide whether to use an electric or

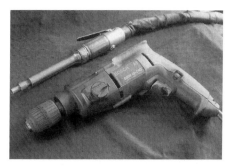

Drill versus air grinder. No contest!

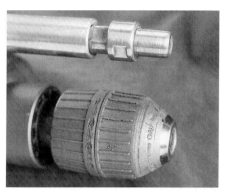

Comparison of grinder collet with modern drill chuck.

air-operated grinder. Whichever you select, it must be of the extended spindle type so the stones and burrs can be used a long way into the ports: choose a 0.236in/6mm collet size.

Electric grinders come in two types - those with the collet chuck mounted on the end of the motor, and those which drive a flexible shaft with a collet chuck handpiece to hold the grinding accessories. The non-flexible shaft grinders are bulky and awkward to use, with access and visibility into the ports being somewhat limited. The flexible shaft grinders are better - in that the body of the motor is away from the workpiece - especially if a variable speed control is available.

Warning! - Electric grinders can be alarming when the stone or burr snags in the port and comes to an abrupt halt. The motor is still running which means the torque tends to twist the grinder out of your hands. Or, in

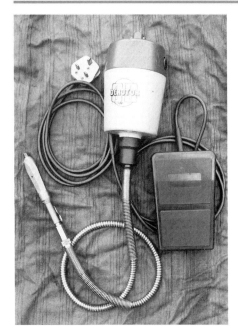

Electric grinder with flexible shaft.

**Air grinders. Standard (top);
0.3hp extended spindle type (middle);
0.9hp heavy-duty (bottom).**

the case of a flexible shaft grinder, the shaft can coil like a snake and then either snap or attack you, so be careful!

For delicate work the authors recommend a variable speed, lightweight, flexishaft grinder with, typically, 120 watts (approx. 0.14hp) at the handpiece, that can spin up to, say, 20,000rpm and hold 0.125in/3mm and 0.25in/6mm shanked accessories.

For heavy duty porting and

chamber roughing out work, we would recommend a 750 watt (approx. 0.87hp) at the handpiece variable speed flexishaft grinder, capable of running up to, say, 28,000rpm and holding 0.25in/6mm shanked accessories.

The authors prefer using air grinders for two main reasons. Firstly, the air motor stalls outright when you snag an accessory in the port, which is much safer. Secondly, cast iron dust is attracted to electric motors, literally like to a magnet, which causes bearings to wear out at an alarming rate and also bridges electrical contacts with a nice crackly short! Running the air compressor away from the dust of the grinding area both prolongs its life and helps keep the noise down.

The recommended type of air grinder is the extended spindle (approx. 5-6in/127-152.4mm) variety where the collet chuck is mounted further away from the air motor by means of a fairly thin shaft (say, 0.75in/19mm on a fairly light grinder). This greatly improves reach and general visibility in and around the ports.

For light to medium use and finer finishing work we use a 0.3hp, variable speed, 28,000rpm, extended spindle grinder weighing 1.17lb/0.53kg. For heavy duty roughing work we have a 0.9hp, variable speed, 22,000rpm grinder weighing 2.75lb/1.25kg.

Typically, a 0.3hp grinder will consume around 9cfm (cubic feet per minute) of air at 100psi. But, if you go and buy a 9cfm compressor, you could be in trouble ... Compressor performance is normally rated in cfm free air delivery (while air tools are rated in actual air consumption). This figure assumes 100% efficiency and is derived from the amount of air the compressor holds multiplied by its

rpm. However, the actual compressor efficiency is only about 75% of this, so a 9cfm compressor would only deliver some 6.75cfm in service. You would need a 12cfm compressor as a minimum, while the extra capacity of a 14cfm compressor would give a good service life when used with the 9cfm grinder.

For enthusiastic DIY work where only a few heads will be modified, the electric grinders are the best bet as they need minimal ancillary equipment. For the compressed air grinders, besides a suitable air supply - which some may already possess for paint spraying - you will need a pressure regulator, a moisture trap (air/water separator) and a lubricant supply (for the grinder's bearings).

A large reservoir tank (receiver) is a must, otherwise the compressor will be running continually, although when you start out there will be plenty of pauses with time for it to catch up, as you check and inspect your progress.

As with the drill, many of the cheaper air grinders have bearings that are not up to the job. Also, with some designs, port access and visibility is little better than with a drill. To save having to continually replace it, get a grinder like professionals use with a long slender nose, a small collet and a good bearing arrangement to support the increased side loading caused by porting. The extra cost is worth it for their longevity.

Also, if grinders operate at a fixed speed, you will find that fine control for grinding is impossible. What you need is a variable speed grinder - you will also find that sometimes the same burr or stone works better at different speeds for different jobs - you can be delicate and do the job just right if you can vary the speed.

Obviously all this kit is going to cost somewhere near the price of a professionally modified head but, if you are at all serious about having a go yourself, it's better to start off with the right equipment rather than struggling and getting frustrated by using tools that are just not up to the job.

PORTING ACCESSORIES

Carbide burrs

By far the greatest rate of stock (metal) removal is achieved by the use of tungsten carbide burrs. There are three basic flute styles, as shown in the accompanying diagram.

We mainly use double cut burrs with a few flame-shaped diamond cut burrs for dressing valve seat inserts.

There are many shapes of carbide burr available. The accompanying table describes the size and shape of the ones the authors commonly use.

When using a burr it needs to be in balance to work smoothly and efficiently. It's best to try refitting the burr into the collet a few times, rotating the burr slightly each time and trying the grinder until the smoothest running is obtained. If the burr still runs rough (possibly due to the burr having been snagged in a port - a not uncommon happening!) a good tip is to gently touch the rotating shank on a fairly thin, sharp-edged piece of metal (eg. the edge of the vice jaws). This leaves a polish mark on the shank where the burr is at its most eccentric. Using a piece of wood or a plastic mallet, lightly tap the burr adjacent to the mark and re-test: by trial and error a slightly bent burr can be corrected to perfection!

Carbide burrs are ideal for working on cast iron and aluminium heads. For aluminium heads use a lubricant such as French chalk, paraffin (kerosene) or WD40 to prevent the flutes clogging.

Shape	Usage	Sizes
Cylinder 	Combustion chamber walls Bathtub chambers	16mm diameter x 25mm long
Ball 	Anywhere a radius is required e.g. around valveguides and chamber walls	8, 13 and 25mm diameter
Oval 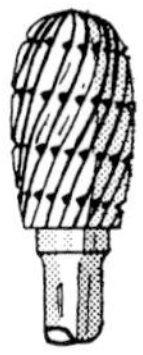	Valve throats, guide bosses and general porting	13mm diameter x 22mm long
Flame 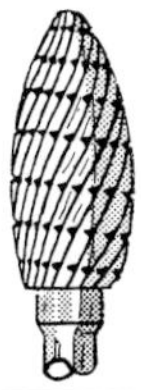	Ports and valve throats	8 mm dia. x 18 long 13 mm dia. x 32 mm long 19 mm dia. 41 mm long
14° Taper Radius 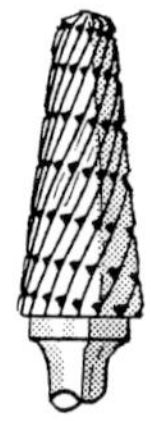 Generous radius 14° Included Angle	Combustion chamber walls	10mm diameter x 20mm long

Various types of carbide burr. All these have 6mm/1/4in shanks).

Mounted points (stones)

The best way of achieving the required shapes on cast iron heads is to employ stones. The material to use is 60 grit aluminium oxide (pink) which can give a high rate of stock removal, or a fine finish, depending on how the stone is dressed. A light pressure dressing gives a fine finish, while a heavy pressure dressing, or tapping the rotating stone,

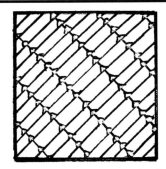

Double cut. The most popular flute style. Very efficient stock removal Creates a small swarf chip.

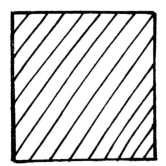

Plain cut. Second most popular style. Drawback is it creates long chips, especially in aluminium.

Diamond cut. Creates a powder-like swarf. Intended for use on heat-treated and tough alloy steels.

Carbide burr cutting surfaces.

gives a coarse finish with a much greater rate of metal removal. Stones **must** be dressed **every** time they are fitted in the collet to ensure they operate smoothly, otherwise vibration causes them to chatter in the port,

Shape	Use	Sizes
W 179: Cylindrical	Reshape as necessary for corners and flats of square or retangular ports, & sides of guide bosses	3/8 in diameter x 1 1/4in long
W 188: Cylindrical	Reshape as necessary for corners and flats of square or retangular ports, & sides of guide bosses	1/2 in diameter x 1 1/2in long
W227: Cylindrical	Roof of combustion chambers and combustion chamber walls	1 1/4in diameter x 1/2 in long
A25: Ball	For radiused ports and for reshaping ports and valve throats to oval	1 in diameter x 1in long

Various stones and their uses. All have 6mm/1/4in shanks.

A selection of aluminium oxide stones and carbide burrs.

Dressing a stone before use.

Roll of aluminium oxide cloth with a split rod.

Flap wheels.

stantly, becoming worse than useless.

The accompanying table shows the basic shapes we use, which we then fettle and reshape to our own requirements.

All of these stones are available from trade suppliers. **Warning!** - Do be careful as the stones from hardware stores may not be safe at high speeds. Always check the speed rating of **all** the grindstones, HSS or carbide cutters you purchase!

Sander bands. Ideal for chamber work.

Discs. Perfect for chamber roof work.

which can do more harm than good.

Stones are cheaper than carbides, easy to reshape to requirement and easier to use - slower stock removal rates make it much more difficult to cause a problem and grind out too much material!

Stones cannot be used on aluminium as they clog up almost in-

Shape	Use	Sizes / grit
Roll of cloth	For using with a split rod to smooth and blend ports	80 grit. 25mm width
Sander bands	Resined cloth band that fits over rubber mandrel. Limited port use but ideal for working chambers	80 grit. 13 mm diameter x 25mm long
Discs	A disc of cloth glued to plastic backing / metal nut. Perfect for flat and curved chamber roofs	60 grit for roughing 80 grit for roughing 120 / grit for finishing
Flapwheel or fan grinder	Narrow strips of cloth resined to shank. Perfect for chamber roofs, walls and round ports	60 / 80 grit. 25mm diameter x 10mm long

Other abrasive tools.

Further accessories

Aluminium oxide cloth, sander bands, discs and flapwheels.

Aluminium oxide coated cloth can be used to finish stoned or carbided ports and chambers.

The accompanying table shows the most common types and their uses.

POLISHING

The authors do not polish their cylinder heads as a mirror finish does not increase horsepower. In some circumstances, a mirror finish can lead to fuel puddling in the inlet manifold/ports, whereas the satin finish the authors prefer generates slight turbulence at the boundary layer, which helps reintroduce the fuel into the passing air and maintain it in suspension. This turbulence may only be a minimal assistance, but any puddled fuel on the port floor will eventually dribble into the chamber where it does not get burned properly. This not only increases fuel consumption, but costs horsepower and gives high exhaust emissions in the form of unburned hydrocarbons (a high percentage HC reading on the tailpipe gas analyser). A satin finish also assists the build-up in the combustion chambers of a thin layer of carbon which acts as an insulator, increasing thermal efficiency.

However, if you wish to polish an item, you can use felt bobs with impregnated waxes or proprietary brass polish or chrome cleaner.

VALVE, SEAT & INSERT MACHINING

Valve seat cutting

Valve seat cutters fall into two categories, those using grindstones to grind the seat and those using metal blades to cut the seat. Which type to use depends on the amount you have to spend when you purchase the equipment and, if purchasing used equipment, what there is available. The following descriptions will give an idea of the pros and cons of each type.

Hand held carbide cutters (*eg.* SP Nu-way)

The cutter uses changeable carbide inserts to cut the seats, together with an expanding pilot to fit the valve guide. The cutters are available pre-set to various angles to enable multi-angle seats to be produced; one cutter does one angle. The diameter of the tool is quite limited so three 45 degree cutters would be needed to cover, say, 1-2in/25.4-50.8mm diameter valve seats.

As the pilots are of the expanding type, seats tend to be very concentric (with practice) and one pilot will cater for quite a range of guide internal diameters. The work rate is very slow, but can be increased with a motorised drive attachment. The really good thing about this system for the amateur is that the kit can be built up bit by bit to produce a comprehensive seat cutting ability. Seat dimensions, vertical and horizontal, have to be measured with depth gauges and calipers.

MIRA-type carbide cutting - magnetic base system

This system uses a single carbide blade with the angles already pre-ground on, so a three-angle seat can be cut in one go. Depth of the seat is measured using the micrometer depth positioning system located on the cutter. Seat width is preset using a valve of the correct diameter to set the cutter. Seats are all identical. A minus point is that the non-fixed pilot (a selection of different diameters is

SP Nu-Way 30, 45 and 60 degree cutters, with expanding pilots and hand tool.

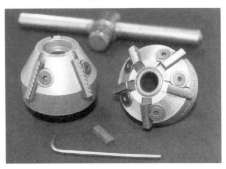

Close-up of cutters showing changeable carbide inserts.

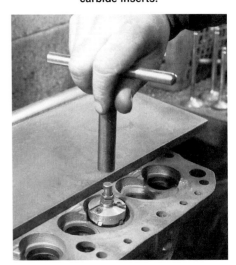

Using the seat cutter.

needed) rotates with the cutter so a clearance must exist between it and the guide. This system allows the seat to be cut non-concentric, not a real problem with practice. The system is expensive but produces rapid,

Selection of 3-angle seat cutter blades.

Cutter setting tool adjusted to valve seat outside diameter.

Valve seat cutting blade adjusted to setting tool to give correct seat diameter.

accurate and identical seats. An optional blade allows the system to be used to machine valve seat insert recesses.

Occasionally, valve seat inserts

Cutting a 3-angle seat with Mira seat cutter.

fitted to some heads contain materials that clog carbide cutters, so the seats have to be ground.

'Primitive' seat grinders

There are quite a few seat grinders of this type, using stones shaped to preset angles to produce the seats. The system uses an expanding pilot to fit the guide, so seats are accurate but widths and depths are **not** very controlled. These systems can be cheap, especially used ones. They are ideal for single-angle seat reconditioning, while multi-angle work is extremely labour intensive.

Grinding can be fairly violent and

Grindstones, expanding pilot and fittings.

'Primitive' seat grinding method.

the stone will chatter as full seat contact is made. A means of dressing the stones accurately to the required angle is needed, together with depth micrometers and calipers for measurement of the seat work.

PEG type orbiting eccentric grinder

The grindstone (preshaped to angle) rotates at around 10000rpm about an eccentric so it is only in contact with a small part of the seat at any one time. The system uses an expanding pilot to position the stone. This system produces very accurate (especially in

Peg seat grinder. Pre-shaped grindstones.

Grinder orbits about an eccentric while high speed stone grinds small area at a time.

Grinder doubles up as a means of dressing stones.

terms of concentricity) seats and is very effective on seat inserts that withstand the best efforts of the carbide cutters. The drawbacks are the need for measuring the width and depth of the seats, changing the stones to vary the angles being ground, and slowness of the work

VALVE REFACING

Used valves, that are in otherwise

This exhaust valve shows damage on seat due to carbon pitting. Over time, its sealing ability becomes compromised.

All is not lost though. If the valve is otherwise in good condition the damaged seat can be reclaimed - in this case by grinding .

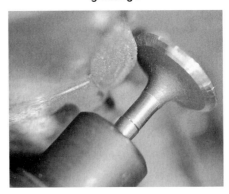

With the seat reclaimed, flow enhancing back cuts can be added.

good condition, can have any minor damage or wear on the seat rectified by refacing. As with seat cutting, valves can be refaced using either cutting or grinding techniques. As valves are made from hard material, the authors prefer grinding.

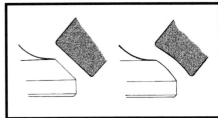

Wear on stone from refacing can cause valve seat problems.

Refacing using cutters
Hand-held cutters are available, the good ones using adjustable carbide inserts to reclaim the seat. The seats produced are accurate, but the work is slow.

Valves can be refaced in a lathe if it is equipped with a taper turning attachment, or is robust enough to use a single-angle cutting tool to cut the full seat width in one operation. Ideally the valve head would be supported to prevent bending during machining.

Producing multi-angle back-cuts is time-consuming by this method.

Valve refacing using grinders
The valve is held in a collet chuck. The chuck rotates and the valve is brought into light contact with a counter-rotating grindstone. The angle of contact between the valve and the stone can be varied to create different back-cuts. Grinding produces accurate, concentric seats with a good surface finish, although the stone requires fairly frequent dressing to maintain it, otherwise rough or curved profile seats can result.

INSERT COUNTERBORE MACHINING

The counterbore, in which a valve seat insert fits, has to be accurately machined to tolerances of approximately one thousandth of an inch (0.0025mm). This accuracy is **vital** to ensure the correct interference

Valve seat insert counterboring tools - multi-manson (right) and IDL.

Insert cutter with pilot in background and a selection of inserts of varying sizes.

Machining recesses for valve seat inserts.

Valve seat insert prior to fitting (left) and fitted (right).

Tools for machining spring seats.

fit between the insert and the head that forms the means of secure location. A metal cutter is required, as stones change dimension with use as they wear.

There are several ways to machine the recess -

MIRA

As mentioned in the seat cutting section, the MIRA type system can be used with an adapted micrometer and a 90 degree blade to machine a head for inserts. The finished cut is very accurate, but time-consuming.

Motorised insert cutters

An example of this type is the Multi-Manson insert cutter. This uses multi-toothed cutters and is positioned by a fixed pilot. Insert recesses are machined quickly and accurately, but set-up time is slow.

Increasing insert I.D. on lathe - easier and more accurate than using carbide or stone with insert *in situ*!

Head Shop Machining Systems

These systems are basically a pillar drill with a universal drive above a flat bed. They can be used to machine heads for inserts, cut seats (using blade type cutters), and bore guide holes out to accept replacement valve guides for those types of head where the guide is an integral part. These systems are very good, being fast, accurate and reliable, but they can be very expensive to buy.

Cutter fitted in guide prior to widening and deepening spring seat.

SPRING SEAT MACHINING

Sometimes the fitted height of the valve spring is incorrect (more detail is given in the valve spring section). If the height is too great, the spring seat can be raised with shims (washers or proprietary shims). If the height is too low, the spring seat has to be machined away. This is done with counterboring tools (flycutters) fitted with at least two carbide cutting tips. The tool can also be used to machine the seats for wider diameter springs.

Smaller versions can be used to reduce the height of the valve guide, should the spring cap foul it or the oil seal. A similar result can be achieved using a large drill bit followed by hand finishing, or using end cutting carbide burrs or sanding discs. A further alternative is to to machine the underside of the valve cap.

Chapter 7
Modification Work

PORTING

When porting cast iron, use the carbides first for most of the major metal removal, then follow with the stones. The process for aluminium heads skips the stoning stage as the material clogs the stone. **Caution!** - Keep the carbide/stone moving **at all times**. Leaving it too long in any one place will create a hollow that will have to be removed. If you're not careful you could end up chasing hollows all around the port and finish up with a

Porting a cylinder head.

mess: moderate pressure on the stone is all that's needed - and keeping it moving. Stop frequently to check progress by running a finger around the port. With a little practice the bumps, hollows and general

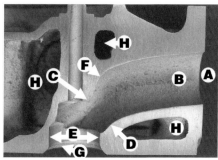

Port terminology. A - manifold face; B - port; C - guide boss; D - short side turn; E - throat; F - long side turn; G - seat; H - waterways.

imperfections will become apparent. The tricky bit is removing/smoothing/blending these without creating further problems for yourself. Work **carefully** and check **frequently**! To begin with,

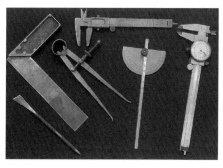

A few of the tools used during head modification work.

it's best to err on the conservative side, by just refining the existing port contours. Aim to remove any obvious imperfections and get a smooth port that is round or rectangular with nice square flat walls, without removing too much metal.

To get the internal dimensions similar for all the ports in the head, a pair of internal calipers is a great help. The ability to judge by feel and sight comes with experience.

To demonstrate how to modify the ports of a cylinder head, we cross-

Photo 1 (see text)

Photo 2 (see text)

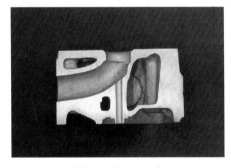

Photo 3 (see text)

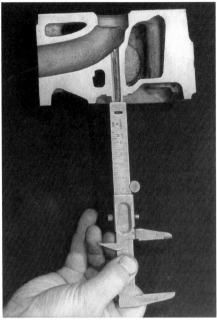

Photo 4 (see text)

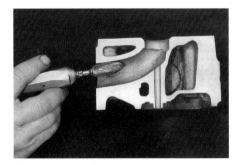

Photo 5 (see text)

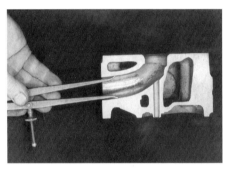

Photo 6 (see text)

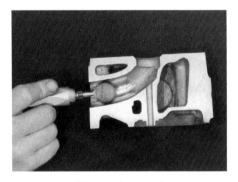

Photo 7 (see text)

sectioned a typical round port cast iron head with integral valve guides - see photo (1). The aim in this example is to remove any obstructions to airflow and slightly enlarge the port.

Removing the guide boss

Using an oval carbide burr (see photo 2) the guide boss is ground away to achieve the shape shown (see photo 3). Work carefully and check frequently, both visually and by touch with your fingers. The extent of

material removal can be checked by measuring the depth of the remaining guide hole (see photo 4). By noting the dimensions of the guide "front and back" the curvature of that section can be duplicated in the other ports.

Port modification with the carbide

The port is rough ground using a flame-shaped carbide burr (see photo 5). The amount of material removed during opening up of the port is checked with the internal calipers (see photo 6). Once you are satisfied with the shape, the dimensions (height and width) at various points within the port length can be noted and the work duplicated in the remaining ports. Scribing lines on the calipers at regular intervals along their length will enable them to be positioned at the same depth in each port.

Stoning the port and guide boss area

Using a large, round ball mounted

point, the port is smoothed to remove any remaining inclusions and marks left by the carbide work (see photo 7). This is continued from the valve end of the port to smooth the area around the guide boss (see photo 8). The end result (see photo 9) is smooth and fairly regular.

Fan grinding port and boss area

The port and guide boss need to be

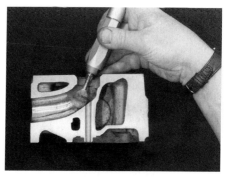

Photo 8 (see text)

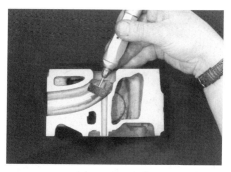

Photo 11 (see text)

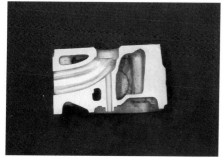

Photo 14 (see text)

Photo 9 (see text)

Photo 12 (see text)

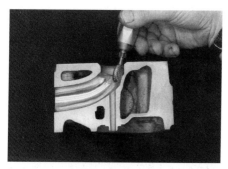

Photo 15 (see text)

Photo 10 (see text)

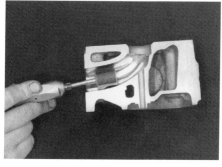

Photo 13 (see text)

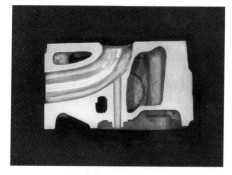

Photo 16 (see text)

fine finished with a flap wheel (fan grinder) to check for any flaws in the stoning process - the fairly polished finish achieved tends to highlight most remaining ripples, bumps and dips (see photos 10 & 11). These are seen more easily and become more apparent when investigating the port with your fingers. The finished port and guide boss area is shown in photo 12.

Flapping the port and boss area

The fan ground surfaces are too highly polished to ensure the air/fuel mix remains homogenous, and there are surface differences where the port has been ground from both ends. This is addressed by flapping the port and guide boss area with emery cloth on a split rod (see photo 13). The finished port is shown in photo 14. The less polished but more uniform finish can be compared to the fan grinder finish seen in photo 12.

Carbiding the valve throat

Having cut the valve seat, the valve throat can be opened up to the required dimensions and blended into the guide boss area. The first stage is to use an oval carbide burr (photo 15). The carbiding joins the finished guide boss and port regions as shown in photo 16.

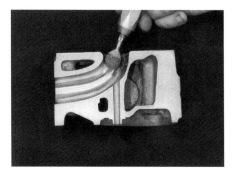

Photo 17 (see text)

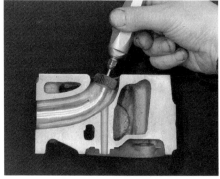

Photo 19 (see text)

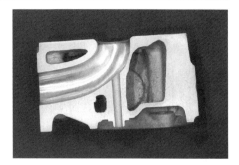

Photo 22 (see text)

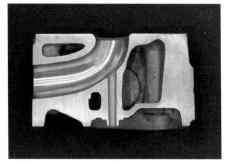

Photo 18 (see text)

Photo 20 (see text)

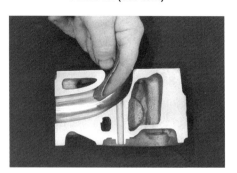

Photo 23 (see text)

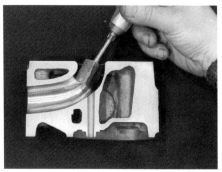

Photo 21 (see text)

Photo 24 (see text)

Stoning the valve throat

The throat is smoothed to remove any inclusions and irregularities from the carbide burr using a suitably shaped mounted point (photos 17 & 18).

Fan grinding the valve throat

The rough finish left after stoning is smoothed and blended using a fan grinder, taking particular care not to damage the transition from seat to port on the short side of the port (photo 19). The polished fan ground finish blends smoothly into the previously flapped port and guide boss area (photo 20).

Flapping the valve throat

The final powered stage of the porting process is shown in photo 21. The throat is smoothed and combined into the port and guide boss area using emery cloth and the split rod. The finished article is shown in photo 22.

Hand finishing

What appears to be straightforward in these photographs is, in reality, anything but. If the cylinder ports were in two halves, the modification process would be simple; however, porting work is carried out fairly blind and often with very restricted access so it is easy to overlook a few minor lumps, bumps, dips and hollows, especially on the short side of the port. Sometimes you can feel the fault but

make the problem worse by not fettling the right area with the grinder. When this happens, it's best to hand finish the short side of the port using emery cloth on your finger. Photos 23 & 24 show this process from both ends of the port.

MATCHING PORTS TO MANIFOLDS

For round ports you can use a gasket as a template, securing it to the head and then scribing the outline,

Race inlet gasket on standard port.

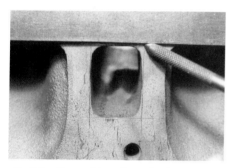

Scribing the outline of required shape on rectangular (exhaust) port.

Open out the corners of the port with a carbide or stone (as here) ...

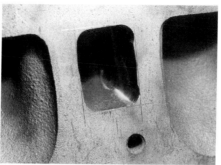

... to get a nice radius out to scribe lines. Not there yet! Getting the walls and floor flat is easier afterwards.

A used valve that has had the margin removed to use as a mask to protect the seats during chamber work (left). A standard valve (right) is shown for comparison.

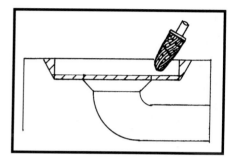

Drawing 1 (see text)

engineers blue applied to the surface before scribing will make things clearer if necessary. With the gasket removed the port can then be opened out with a carbide to within the markings and blended as the rest of the roughing work in the port is done. Finishing and blending to just within the scribe lines can continue as the stages covered previously are followed.

For a rectangular port you can use either a gasket as a template or an engineers square to scribe the straight lines to outline the required port shape. Open out the corners first to a nice radius, either with a small flame-shaped carbide or a cylindrical stone. You should aim to get the port walls flat, using a flame-shaped carbide at first, then a cylindrical stone. Modifying the throat and guide boss area (with suitably sized stones) and the fan grinding and flapping are as shown and described for the round port.

COMBUSTION CHAMBERS - RESHAPING & FINISHING

Combustion chambers are more difficult to modify and finish than ports. A slip can ruin the head - for example, grinding back too far on the chamber wall can undercut the gasket line; rectification would involve welding, an expensive and often not successful process. Photographing a cross-section of a chamber might not show clearly the areas to be modified, so we have used rather exaggerated line drawings to illustrate the techniques involved.

Chamber wall shrouding
Using templates and a gasket to mark out the metal to be removed, use an appropriate carbide burr (we usually use one with a 14 degree included angle for consistency) to fettle the walls. Be careful not to sink into the chamber roof or you might dip below the valve seat level - not cosmetically appealing ! This process is shown in drawing 1.

Chamber roof roughing
Drawing 2 shows the chamber walls cut back and the stone used for rapid stock removal to bring the chamber roof flush with the top of the valve seat. This stage can be done with a cylindrical carbide burr on aluminium

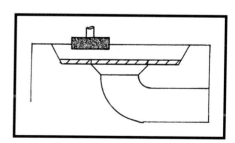

Drawing 2 (see text)

Using a cylindrical stone to modify the chamber roof.

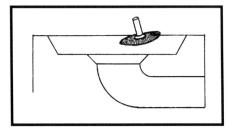

Drawing 3 (see text)

heads. A pair of used valves that have had the margins removed are used as masks to avoid damaging the valve seats. Aim to achieve a smooth finish or final finishing will be time consuming and expensive in materials.

Chamber roof finishing

Drawing 3 shows the use of a sanding disc to smooth the roof - be careful where the roof meets the walls as the disc can cut into the wall, damaging it.

You can also use a fan grinder to

Finishing the chamber roof with a sanding disc.

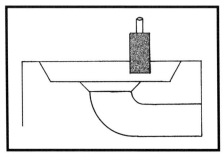

Drawing 4 (see text)

fettle the roof and blend the chamber walls at the same time, especially with aluminium.

Chamber wall finishing

The wall can be fine fettled using a sanderband and possibly a fan grinder for a nice smooth finish. On Hemi-heads it can be difficult to achieve a uniform finish, so sometimes we use emery cloth on a split rod and carefully smooth and blend the chamber. Try not to overlap the head face - it is advisable to complete this

process before skimming! Drawing 4 shows this process.

MATCHING MANIFOLDS TO GASKETS/PORTS

The gasket to be used on the engine makes an excellent template for marking out the regions to be opened out. Most inlet manifolds are aluminium so use a carbide (dipped in or sprayed with some WD40, or similar, to reduce clogging) for fettling, followed by emery on a split rod.

The idea is to not open out the manifold all the way to the marked line but, instead, to leave around 0.020in/0.5mm of material all the way round. When the manifold is finally fitted to the head, it may not align accurately with the ports, in most cases you cannot see this. The step will allow for any small shift of manifold position and is in the right direction in flow terms; from manifold to port and not from port to manifold. The step also creates a slight turbulence which picks up fuel droplets and helps keep the fuel in suspension. It also allows room for the gasket to exude slightly as it is compressed without impeding airflow.

With most sidedraught manifolds, especially for race applications, you can mount them on the head with the gasket and fettle inside until they exactly match the ports in the head. However, while some people prefer to match the exhaust port to the manifold exactly, we feel that a step from the port to a larger diameter manifold, say, 40 thou'/1mm all the way around, discourages sound waves and seems to help the engine come on cam earlier.

Chapter 8
Valves, Blueprinting, Buretting, CR & 'Unmodified' Class Heads

CHECKING USED VALVES

You'll need to establish whether the valves are worn and in need of replacement. First, check all the valve stems for wear and damage, such as ridges or scoring. You can either use a micrometer for the job, or try mother nature's own test equipment - available right at your fingertips. By lightly running finger and thumb along the length of the valve stem, and working all the way around it, any slight ridges should become apparent (you can feel wear of a thou' or less). If **any** wear can be felt, new valves are in order. The alternative is to measure the valve stems with a micrometer. Check in several places around the stem, concentrating mainly on the region that actually runs in the guide. Compare this with the measurements in the workshop manual; again, if there is **any** wear, you need to get new valves.

You can check stem straightness by rolling each valve stem on a a flat surface and looking for gaps between the stem and the surface.

Don't throw the old valves away,

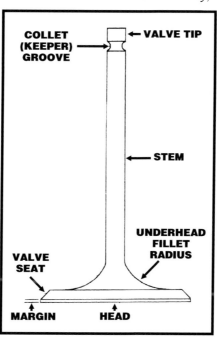

COLLET (KEEPER) GROOVE ← VALVE TIP

← STEM

UNDERHEAD FILLET RADIUS

VALVE SEAT

MARGIN   HEAD

Valve anatomy.

they'll come in useful later as masks for the seats when chamber modifying.

CHOICE OF VALVES

Now would also be a good time to consider the level of modification that you'd like to undertake with the head, keeping in mind any other intended changes to the engine.

If the engine specification is to remain pretty much as standard, if you plan to retain the standard cam profile or only run a mild high performance cam in the engine, and if you don't use the engine much over 5000rpm, or so, there's no need to change the valves to ones of increased diameter. Retaining the standard valve size will give the engine improved flexibility and performance right where you need it, through more low and mid-range power.

Inlet valves
The inlet valves fitted as standard to

Inlet valves - standard (left); back cut (middle); custom (right).

Close-up of standard and back cut inlet valves. The 30° back cut removes the lip that is visible on the standard valve, improving flow markedly.

Re-shaping and polishing the back of a valve to improve airflow.

most heads are made from EN52 steel which has good wear resistance and is perfect for the job; in fact, some lowly stressed engines use EN52 for the exhaust valves, too. Stainless steel is best for the inlet valves of racers).

Going larger with the inlet valve sizes is really only necessary for rally or race applications, to give large bore road engines real performance and

sparkle (they can seem rather 'flat' otherwise), or if you wish the ultimate road performance with a smaller engine (combined with performance cams and further engine uprating). Bigger valves mean a great deal more head modification work is necessary to allow the valves to achieve anywhere near their flow potential. With bigger inlet valves, smaller bore engines will lose some of their low rpm tractability - more noticeably than bigger bore engines - as a trade-off for more top end power. It also means more expense when compared to purchasing standard replacement valves, as the larger valves will have been custom made, from either the EN52 steel or 21/4N stainless, and that doesn't come cheap.

Exhaust valves

The specification of material used for the standard exhaust valves in most cylinder heads is 21/4N Austenitic Stainless steel with a chrome plated stem and a hard Stellite tip (providing greater wear resistance) for where the rocker pad contacts the valve. The 'new wave' treatment is to tuftride the valve stems as an alternative to the chrome plating. This is a chemical process that leaves the stem with a hardened wear resisting surface without actually altering its diameter by depositing a layer on it. This gives a stem finish with none of the high spots or the sharp bits that can arise from chrome plating. 21/4N is the best material for the job, so there is no need to change to another fancy grade of material: such valves are already well capable of withstanding the rigors of unleaded fuel. Some heads that have already been reconditioned may have had new exhaust valves of a lesser material specification fitted to replace existing ones if they were badly worn or damaged in some way.

One problem arising from worn oil seals/guides is carbon build-up on the backs of the valves ...

... which is detrimental to airflow - think of the restriction this would be causing! Repair/replacement of guides and seals with a thorough de-coke would restore lost sparkle and improve mpg!

If the existing valves from the head are not too worn, and you are intending to re-use them, a quick test of the exhaust valve material is whether or not they stick to a magnet. The Austenitic stainless used for valves is non-magnetic so, if the magnet sticks, the valves will need changing. Many manufacturers' exhaust valves are now of a bi-metal construction, having wear resistant EN52 stems (magnetic) friction welded to a 21/4N valve head (non-magnetic), this is all right for most road applications where standard revs are not exceeded.

Steel 'bristle' valve guide cleaning rotary 'files'. Ideal for removing the sticky carbon deposits associated with lead-free fuel composition ... failure to clean can cause valve to guide seizure.

The authors feel that, for road use, it is unnecessary to fit exhaust valves that are of a larger diameter than standard, and we've found this to be true even in our championship-winning full race heads. So, if standard exhaust valve size works well, even in the case of race engines which operate at rpm levels far higher than road engines, we see no need to go any larger. This statement falls a bit flat in the case of turbocharged or supercharged engines, but we'll come to that a bit later.

It might be useful to offer an explanation of the thinking behind favouring the standard size exhaust valve. It is based on the pressure differences that exist in the cylinder at the end of the power stroke/beginning of the exhaust stroke, and the area of valve curtain available for the exhaust gas to flow through. After the work has been extracted from the fuel by the piston being pushed down the bore, the residual gas in the cylinder is still at a considerably higher pressure than the pressure existing in the exhaust manifold and exhaust system. When the exhaust valve opens, the gas leaves the cylinder at very high speed due to this large pressure difference (called "exhaust blowdown").

For example, say the residual pressure in the cylinder when the exhaust valve opens is roughly 70psi. Using a spot of rule of thumb maths, the flow rate is based upon the square root of the pressure difference across the valve. Assuming the exhaust manifold is at atmospheric pressure (14.7psi), the flow rate difference to the inlet valve is $\sqrt{(70/14.7)} = 2.18$. This means the exhaust will flow just over twice as much as an equivalent sized inlet valve. It's far easier to get the used stuff out than the fresh stuff in!

You can argue that a small valve will cause a bottleneck for the waste gas trying to get out. But, as most cams begin to open the exhaust valve well before the piston reaches BDC, any residual pressure is going to continue to contribute to pushing the piston down toward BDC, whilst the gas is still trying to get out of the exhaust port, hence generating more torque.

Any remaining exhaust gas is then expelled by the upward motion of the piston once past BDC.

The egress of the gas is further aided by the exhaust port and seat acting like a funnel, with the flow being from a large space (cylinder) to a small space (exhaust port). As a funnel is very efficient at flowing air in this direction, the exhaust gas leaves the cylinder very rapidly, and modifications to the exhaust valve, seat and port further aid this process. With the exhaust gas leaving the cylinder so rapidly, there can be a partial vacuum left behind in the cylinder afterwards. This reduced pressure then helps with drawing in the fresh mixture on the following induction stroke. However, the amount of help this lower pressure generates depends on a myriad of other factors, such as engine rpm and camshaft and exhaust manifold design.

Exhaust blowdown takes a certain length of time to occur, irrespective of the speed (rpm) of the engine. But increased engine speed (rpm) means the exhaust valve is physically open for less time. So, above a certain rpm, complete blowdown cannot occur as there is insufficient valve open time for the volume of exhaust gas to be expelled. Some exhaust gas is left in the cylinder. This remaining exhaust gas, besides occupying some of the space in the cylinder that should be filled by the incoming charge of fresh mixture, also contaminates the fresh mixture, and power is lost (deliberate contamination with exhaust gas is used as a means of reducing emissions). Some of the remedies for this involve increasing the valve area, opening the valve sooner (earlier), higher (more lift), longer (more duration), or combinations of all four. The first solution is achieved by fitting a larger exhaust valve, giving more area (a bigger hole) for the exhaust gas to flow through. The latter three, however, come under the control of the camshaft and valvetrain.

In the majority of cases, for road use a larger than standard exhaust valve is unnecessary. As most modified or race engines have a performance cam fitted, any problem of poor cylinder emptying cannot readily occur.

There may be a flaw somewhere in this reasoning, but it works for us!

Forced induction

As previously mentioned, forced induction changes matters somewhat. The extra air and fuel mix that is forcibly crammed into the cylinders means that the exhaust valve can be hard pressed to cope with the extra volumes of gas it has to handle. In such cases it is best to go for the biggest exhaust valve that will fit, even to the extent of retaining standard size inlet valves, or sacrificing some of an inlet valve size increase to allow for the fitment of bigger exhausts.

With turbocharged engines, getting the gas out and away as fast as possible will help spool up the turbo and make boost, while supercharging just requires help with emptying the overfilled cylinders. As a rule of thumb, we size exhaust valves at 90% of the inlet valve size for forced induction applications.

VALVE GUIDES

When replacing valve guides there are several alternatives available. For cast iron heads the choice is between cast iron or bronze alloy replacements, whilst the majority of aluminium heads have bronze alloy guides as standard. In both instances, another option is to use thin bronze guide liners (K-Line inserts). New guides are used where the originals can be easily removed or, where the valves run directly in the

material of the head without a separate guide as such, the existing holes can be bored out sufficiently to allow the fitting of guides. K-Line inserts, or equivalent, are a very effective way of reclaiming worn or oversized guides that may be difficult to recondition by other means - provided the originals are not worn beyond reclamation by this method. The existing guide hole is reamed out oversize and the liner, in the form of a thin bronze tube, is fitted. It is then expanded to fit the guide by the use of three successively larger broaching tools, hammered through one after the other. The liner is securely located by conforming to the shape of the guide. Any surplus tube left protruding is topped and tailed, and the liner is ready to be reamed out to size for final running clearance.

Note that, for competition use, valve guides that are integral with the head are best bored out to take press-fit bronze guides.

Standard cast iron guides are fine for use with a standard head, but as we're interested in improving the head's performance, bronze guides are definitely the ones to go for. This is especially true if unleaded fuel is going to be used. Bronze guides offer a far greater ability to transfer heat away from the valve stem, have better wear

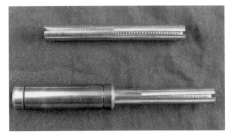

K-line thin bronze guide liners, showing one part fitted into guide.

K-line installing equipment, reamer and pilot (top) with the three broaching tools near the bottom. The topping and tailing tool is on the right with a pile of inserts (left).

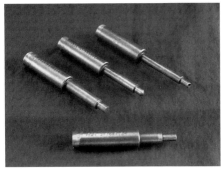

Various punches for removing guides from those heads where it is possible to do so ...

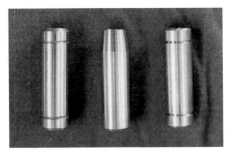

Not that it's all that obvious in black and white ... standard cast iron valve guide (left); bulleted bronze (middle); standard bronze (right).

characteristics and can run with reduced lubrication compared to the cast iron variety. Bronze guides also run tighter clearances, so less oil gets down the guide to contaminate the incoming mixture, or carbon up the back of the valve. After many years of research and development, we now

...using the time-honoured method of vibratory and percussive persuasion ... namely a large hammer!

use our own special Manganese Silicon Bronze alloy valve guides in all our heads. Made from a material specifically developed for use as valve guides, these guides are suitable for running treated, chrome stem or non-chrome stem valves equally well, unlike some types of bronze guides that can gall and wear badly if chrome/treated stem valves are not used. Always ask a supplier what valves should be run with the bronze guides they are selling to avoid any wear problems later. If in any doubt, get valves with chrome stems - inlets and exhausts.

It is the ability to transfer heat away that is the most important aspect of using these bronze guides but their ability to cope with unleaded petrol is also important with regard to the way modern fuels are developing. With the reduction in the amounts of lead in fuel (eg. the UK once had 0.4g lead/litre but from 1986 it became 0.15g/litre leaded, 0.013g/litre unleaded - and will be reduced further in the future, no doubt), the petrol companies are increasingly turning to other additives to try and replace the effect of lead. In case you didn't know, lead - or, rather, tetraethyl lead - as well as lubricating the valve and seat in the head, acts as a flame retardant and slows down the rate (speed) of

burning of the mixture in the cylinder. Without the lead, the fuel burns much faster and hotter (hence the need to retard ignition timing for cheap unleaded fuel - not super unleaded, though - to compensate for this) giving the valves a hard time. The exhaust valve has to endure by far the worst conditions, given the extreme heat from the burnt gas passing through on its way out of the engine, as well as the continual pounding against its seat in the head. The inlet has it somewhat easier, being cooled by the lower temperature fresh mixture passing around it as it's drawn into the cylinder.

Bronze guides do cost more than the standard cast iron replacement variety, but there is no point in skimping when it comes to modifying your head; fitting them is definitely to be recommended. They come as straight bronze tubes, which are fitted by pressing them into the head and are then reamed to give the correct valve clearance. Reaming after fitting is **essential** to remove any distortion that has occurred during fitting and to guarantee correct running clearance.

Caution! - As for knurling the inside of worn guides as a means of reclamation, **don't**. Enough said!

When the guides are for use in a modified head you may wish to get them 'bulleted' before fitting, which helps to reduce the obstruction caused by the guides where they protrude into the port. The guides should be machined with a 7° taper from one end, the exact length of this taper depending on how far the guide protrudes into the port.

It is by far the best idea to carry out some of the grinding work in the ports and on the guide bosses before the guides are fitted and the seats are cut. This way there is no chance of accidentally damaging the new seats;

Pressing new guides into a head.

After fitting it is essential to ream out the guides to correct any distortion or compression caused during installation ...

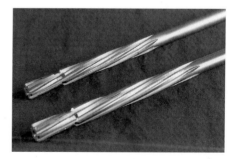

... using reamers such as these of just one thou' difference in diameter - spot the larger one.

working in the ports, too, is a little easier without this worry. Hitting a seat with a carbide cutter or a stone will mean the seat having to be re-cut, which may cause problems later when balancing chamber volumes, not to mention costing more money.

VALVE SEAT INSERTS & CONVERSION TO UNLEADED FUEL

Whether the head was found to be cracked around the valve seats or not, we would strongly recommend having hardened inserts fitted in place of the standard exhaust valve seats for cast iron heads. The age of most of the cylinder heads which are being modified, as well as the hard life they lead, usually means that some heads will already have been reconditioned at least once and are likely to have had the valve seats re-cut. Those older heads that have not been reconditioned will be in poor condition due to exhaust valve seat erosion and damage caused by today's low lead fuel. Cutting new seats in the head will cause them to sink further into the chamber, which means the valve sits

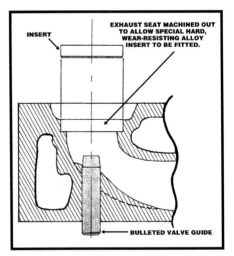

Cutaway showing fitment of a valve seat insert.

higher in the head and the valve train geometry can be upset. Valve seat inserts will act as protection against increased wear, running equally happily on leaded or unleaded fuel - though you may as well run the less expensive unleaded fuels. This way you save money, so the conversion will pay for itself!

VALVE SEATS

On the subject of valve seats, the only ones to have on any head, performance or otherwise, are what are called three-angle seats (see drawing). These are made up of three angles as the name suggests, a 60° bottom cut that blends the valve throat into the 45° valve seat, followed by a 30° top cut that merges the valve seat into the chamber. It is this region of the cylinder head where the greatest gains in airflow can be realised because, as mentioned before, airflow hates abrupt changes in direction or volume, and the flow from the port past the valve offers one of the greatest obstructions to flow encountered in a head. Anything that can be done to smooth the transition of the air from the port to the chamber reaps substantial benefits. In the case of three-angle seats against once conventional (45°) single-angle seats, the benefit realised in favour of three angles can be an immediate 25% increase in flow! So, without these high flow three-angle seats, you're behind in the game already - it takes some mighty fancy porting work to match these flow gains with a single-angle seat, and you pay for the time taken to do it, too! In some cases - certain classes of racing, for instance, such porting work would not be allowed.

The new seats should be cut so that the valve seals around the outer

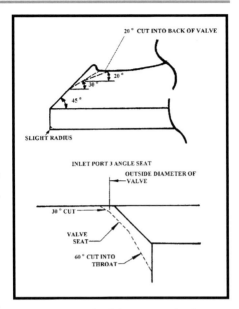

3-angle cut for inlet seat and valve.

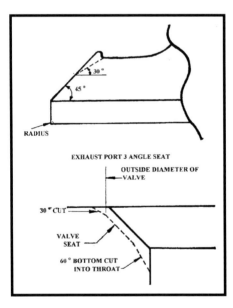

3-angle cut for exhaust seat and valve.

edge of the valve seat and not the middle or inside edge. This increases the valve seat diameter and therefore maximises the width of the port throat in the head by producing a wide 60° bottom cut. When modifying the head this broad 60° bottom cut can be reduced in width by grinding out the throat with a stone, so increasing the

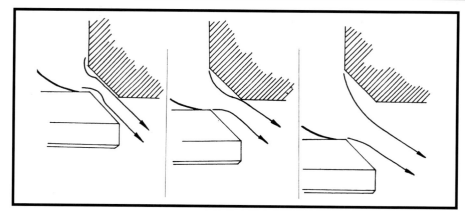

Pattern of airflow past a standard seat and valve for low, mid and high lift.

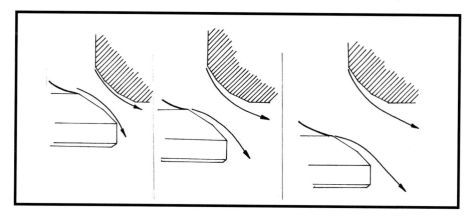

Pattern of flow through modified seat and valve for low, mid and high lift.

BLUEPRINTING & BURETTING

Blueprinting.

We can now get down to describing the modifications that can be done to the head.

In this chapter we'll cover the, apparently, most straightforward cylinder head - though this is quite far from the truth, technically speaking - the 'blueprinted' standard head used in some forms of competition.

These 'standard' heads are used on engines for cars competing in standard "unmodified" motorsport classes. The regulations for these classes usually state that absolutely no modification to the ports or valves is permitted, so no porting or polishing is allowed whatsoever. They may, however, state that valve seat specification is free and conversion to run unleaded fuel is allowed (often in classic car racing classes). Having the lead free conversion allows the use of higher octane super unleaded petrol, which means the engine will take more ignition advance and make more power.

The "no modifications" regulation means that a specially selected casting is necessary in order to comply with these rules, and yet still allow the engine to make race-winning power. If more than one type of head is available, it will be necessary to refer to class rules as to which versions are allowable, hopefully those which come with larger inlet valves as standard. The head also needs to have the absolute minimum of casting faults and/or core shift visible, so the throats of the ports have no misalignment with the valves, or rough misshapen walls. It is also vital that the valve seats are not badly damaged or sunken in the chambers, as cutting new valve seats would leave nasty sharp edges in the

diameter of the throat and allowing the creation of a better shape for the airflow to follow. The bottom cut serves as an excellent visual aid for grinding work, acting as a guide to enable the throat to be opened out evenly all round, and providing a nice visible lead-in before getting close to the seat itself.

There are a variety of different seat cutter widths available, the most common producing 1.5mm/0.059in seats, 1.3mm/0.051in seats or a 1mm/0.039in seats (the latter for race use). The majority of seats for heads intended for road use are 1.3mm wide, a big difference compared to the standard head's usually wide seats. The 1.3mm seat width used is

sufficient to allow the essential heat transfer to take place between the valve and the cylinder head when the valve is closed, to ensure mechanical reliability and longevity of the seat, yet will allow improved airflow through a wider throat.

The seats should be cut in such a way that all the valve heads finish up at the same height, within a tolerance of plus or minus two thou' (0.002in/0.05mm). If the valve heads sit at differing heights in the chambers, this will have a marked effect on the chamber capacities, making life difficult when it comes to equalising the chamber volumes in order to balance the compression ratios of the individual cylinders.

chambers. If the rules state "no chamber modification" these sharp edges cannot be removed afterwards and, besides interfering with the airflow, such edges will create hot spots in the chambers, causing damaging engine pinking or worse, destructive detonation.

Needless to say, these physically perfect castings are thin on the ground and finding a suitable one takes a lot of searching, not to mention the experience necessary to know what to look for in the first place, even if you're fortunate enough to have a stock of old heads to look through. So you will need to consult a reputable specialist tuner for your particular engine. Once a suitable casting has been found, it is then cleaned up and tested on the flowbench to see if the head actually performs up to par (or, preferably, above) in terms of airflow. The ports may look okay but, as we've said, looks can be deceptive, and flow testing is the only way to sort the wheat from the chaff.

Once the final choice of casting has been made, new guides are fitted - bronze alloy for the exhausts, standard cast iron for the inlets. Valve seat inserts are fitted for the exhausts (the bronze guides and inserts are used as part of the permissible lead free conversion, if that's part of the regulations). High flow three-angle valve seats are cut for both the inlets and exhausts. Seat widths are usually 1.3mm, with a minimum top and bottom cut. Once the seats are cut, any sharp edges remaining in the throat after the machining must be carefully removed by hand using a fine file.

Clean out the guides with some spray choke/carburettor cleaner followed by pushing a small wad of tissue paper or rag through the guide to remove any swarf that may cause

Lapping in the valves with some fine lapping compound ...

... to give an even grey seat all around the edge of the valve.

damage to the valve or guide itself. All that is required now is for the valves to be lapped in with some fine lapping compound. Only a few spins of the 'sucky stick' should be needed to give an even, continuous grey seat around the outer edge of the valve and on the head. Keep together the pairs of valves for each chamber, they will be required to find the chamber volumes.

Usually this is as far as the rules

say you're allowed to go with heads, except for skimming the head face to get the desired compression ratio (compression ratio or chamber volume may be specified in class rules). As to the amount of metal removed by skimming, it's impossible to specify that quantity here: the irregular shape of the combustion chambers means that no two heads are alike, and it is not possible to give an exact amount to skim off to achieve the desired chamber volume. Measuring each combustion chamber's volume by 'buretting' is necessary.

Buretting.

To measure the chamber volumes you will need a burette, a piece of perspex large enough to cover the chamber and with a small hole drilled in it, and some grease. You'll also need a spark plug of the specific type to be used in the engine: this is **essential** as the spark plug's shape and design can make a difference to the volume readings. Install the spark plug firmly into its hole, with a very light dab of grease on the threads before installation to ensure a good seal. Place the head on some form of stand - a couple of wooden blocks will do - with the chambers uppermost. Lightly grease the inlet and exhaust valves around the seat and install them in the chamber to be measured. Then put a thin smear of grease on the head face around the outside of the chamber and stick the perspex onto the head. Don't allow too much grease to escape from around the valves or from under the perspex into the chamber. The quantity of grease will obviously detract from the measured volume. Fill the burette with water and zero the reading on the tube. You can use alternatives to water - paraffin or similar, but it demands far more care

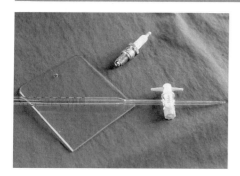

Buretting equipment - not including the blob of grease!

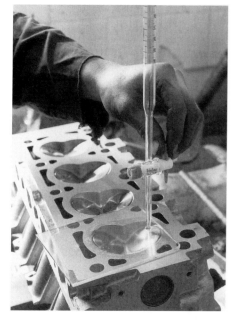

Measuring the chamber volume.

Measuring chamber depth to calculate how much to skim from a head (see Appendix).

Skimming a head on a milling machine using a flycutter.

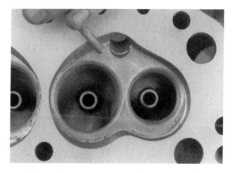

Deburring the edges of a combustion chamber with a fine file.

repellent, to stop it going rusty. You will be surprised how fast corrosion strikes!

Repeat the process for the other chambers, remembering to use the actual valves that were lapped for those chambers, and making a note of each chamber's volume as you go.

You can then use these measurements as a guide to how much material has to be skimmed from the head.

You will undoubtedly find that one of the combustion chambers is slightly larger than the others. This is a common occurrence on many heads and, for our blueprint specification heads, it is this chamber that, by skimming, we get as near as possible to the volume given in the regulations. Several further checks of the volume of this larger chamber will be necessary, using the burette, between skims. These repeated checks are essential in order to avoid removing too much material and ending up with too small a chamber. With the largest chamber reduced to the required size, the other three chambers will be too small, their volume now being slightly under the regulation size. As modification of the chambers is not allowed, the thickness of the inlet valves in each chamber is reduced by re-facing the seat, so making the valve margins thinner. Most inlets have very thick margins as standard, and thinning them in this way does not compromise strength or longevity. More importantly, the valve face remains unaltered. This requires repeated reassembly and measuring of each of the chamber volumes with the burette until all the chambers are equal to within 0.5cc. The valves are then number stamped so they can be replaced in the correct chamber if the head is ever disassembled at a later date.

in use, is not very kind to your hands and, besides, water is more readily available and spills don't matter. You can now measure the chamber's volume by carefully running the water from the burette into the chamber through the hole whilst allowing the displaced air to escape. Once the chamber is full, with any lingering air bubbles removed, the burette will indicate the chamber's volume as the amount of water that has been used. Make a note of this reading. Empty the water from the chamber and give it a quick squirt of WD40 or similar water

All this may seem a bit extreme, but is **vital** if an engine is going to run smoothly and produce race-winning performance. Standard class-type motorsport is intended to equalise the level of modification allowed to the engines, and so give good closely matched competition. If you want to win in such classes you have to exploit any legal means of improving performance, and this apparently manic attention to detail is crucial.

The final process is to run a deburring tool or a fine round file around the edge of each of the chambers to remove the nasty sharp frazes left by the fly cutter used for skimming. The idea is to do the minimum of metal removal conducive to breaking those sharp edges, to remove the chance of any hot spots forming as mentioned before. A quick deburr of the outer edges of the head face would not go amiss, either, to save your hands from the sharp saw tooth finish when carrying the head around.

As a brief aside, a mention of the method of skimming the head would probably be beneficial here. The best technique, by far, for metal removal from the head face is using a fly cutter. This removes the material in a series of circular sweeps as it traverses the length of the head, leaving distinctive curved machining lines visible on the head face once skimming is complete. This creates an excellent flat surface, but one that has a rough finish. The coarse surface finish is essential, it enables the head gasket material to conform and compress into the tiny grooves, vastly improving the gasket's sealing ability and performance under combustion and other pressures. Skimming the head by any other means does not allow the gasket to 'key' into the head. Obviously, the smoother the head's finish the poorer the seal will be, with the resulting water leaks and eventual loss of power due to a blown gasket.

COMPRESSION RATIOS

For road use it is probably best to stick to a maximum of 9.75:1. The low octane ratios of modern fuels makes going higher impractical. For rally and sprint use the CR can be raised to 10.8:1 and for race use 11.5 to 12.5:1 (the large amount of overlap from the cams used serves to reduce the cylinder pressures).

These compression ratios are merely guides and should not be taken as gospel. If the compression ratio is increased much beyond these recommendations, pinking and even severe detonation can occur. To stop these problems the ignition timing would have to be retarded, and the result of not having enough ignition advance can be a hefty loss of horsepower! Details of how to calculate the optimum compression ratio for your application are given in the appendices.

Chapter 9
Valve Train Components

VALVE SPRINGS

The valve springs provide a positive means of closing the valve and keeping the follower in contact with the camshaft lobe so the motion of the valve train is controlled.

This fairly straightforward description is a massive understatement when it comes to covering the complex and vital role that valve springs play in the process of making horsepower.

It is the job of the cam lobe and the valve spring to control the motion of the combined mass of the valve, the cam follower or bucket, the pushrod (if used), spring retainer (spring cap) and collets (keepers), as well as the mass of the moving part of the spring itself.

As described in the camshaft chapter, the cam lobe provides the mechanical means of converting the rotary motion of the camshaft into the reciprocating (up and down) motion of the valve. The lobe profile dictates the

time and speed of valve opening, either via the follower, pushrod and rocker arm, or by more direct action on a follower or bucket in the case of overhead cam engines. All well and good, except that there is no direct link between the cam and the valve to pull it shut again. Valve opening and closing corresponds to the regions of the cam lobes known as the clearance ramp, the opening and closing flanks (or ramps) and the nose. The lobe pushes on, and accelerates, the entire valve train during the positive acceleration period of the opening flank. This is the only time the valve spring does not have to work to keep everything in contact. During the negative acceleration (slowing down) period of the opening flank, the entire valve train would continue on regardless at its maximum velocity, were it not for the valve spring. The springs must be strong enough (have enough pressure) to keep all the valve train parts, and hence the follower, in

contact with the cam lobe at all rpm up to the engine's operating limit, ensuring the whole valve train is brought to rest at full lift (cam nose), so valve float does not occur. They must then ensure that the cam profile is followed away from full lift back to the valve closure position.

Valve float

The higher the rpm the engine is turned to the greater the force the cam imparts on the valve train due to acceleration. As a result, the mass of the valve and its components becomes more and more reluctant to slow down or change direction (it possesses greater inertia). There comes a point where the valve spring is no longer of sufficient strength (force or open pressure) to combat the valve train inertia and valve float occurs. The follower is literally flung away from the cam lobe by the acceleration rate of the opening flank, momentarily losing contact with it. Of course, during this momentary separation the cam has continued to rotate, whipping round further, so when the follower lands back on the lobe it has missed a portion of the profile and its motion is no longer quite so controlled.

Valve float obviously messes up the timing of the inlet and exhaust processes during the four stroke cycle. It manifests itself as a sudden dramatic loss of power at whatever rpm it occurs, similar to a severe engine misfire. Mechanically it is not good for valve train longevity; everything crashing back together can bend or break things, like pushrods or rocker arms, not to mention damaging the cam lobe itself.

Valve float can be brought on by several factors. Over-revving the engine, tired or worn out valve springs

New spring (right) compared to worn (left). Note the difference in height.

(they lose their resilience over time due to heat and because of the very nature of their movement), or too aggressive a cam profile can all produce this same result. The solution is to use new or stiffer springs.

Valve bounce
Valve bounce happens when the valve hits its seat with sufficient force to bounce off again - potentially disastrous if the piston is in close proximity! It is down to the vibrations

Professional valve spring tester. In drag racing, checking spring poundage is critical.

in the entire valve train, when the engine is at high speed, acting together for an instant so that the valve arrives back at its seat much faster than intended, as it is no longer under the control of the cam or valve spring.

The cure is to stiffen up the valve train components, making them less likely to flex. Stiffer pushrods, stud girdles, stronger rocker shafts and rocker posts, etc, will help to alleviate this problem or, at least, move it to an rpm level beyond the engine's operating capability. The components should be braced or stiffened, preferably without increasing the mass of the moving parts which could lead to valve float, and you'd be back at square one.

Spring surge
A helical valve spring has a frequency at which it will resonate (like a tuning fork), called its "natural frequency." This frequency is to some extent

dependent upon the design and construction of the spring. It will also vibrate at multiples of that natural frequency.

The disturbing forces that cause spring surge are related to the design of the cam's opening flank and the speed or rpm of an engine. At a high enough rpm the frequency of the valve motion can cause the valve spring to resonate. It starts to 'do its own thing' in terms of motion, which inevitably means that the spring's effective pressure is reduced dramatically, or possibly eliminated altogether, meaning the valve train motion is no longer controlled. This may have a catastrophic effect on engine longevity.

To counteract valve float and spring surge, stiffer springs are used. To make a spring stiffer the design must be changed. The wire for the coils could be made thicker, the type of material for the wire changed or the diameter of the spring altered. This is fine up to a certain point, but problems can arise from the spring becoming coil bound at maximum valve lift, or no longer fitting physically within the constraints of the standard head, meaning machine work or a total change of valve train to accommodate.

As the need for stiffer springs increases, it can outstrip the capability of a single valve spring. The next step is to use double springs and even triples in the case of high rpm race engines with radical cams. With doubles, a smaller (diameter and wire thickness) but taller spring, wound opposite to the main, is fitted inside. It is wound the opposite way to stop the two sets of coils tangling should the inner one move. The different natural frequency of the smaller spring can help cancel out any operational problems with the natural frequency of

Double valve springs.

Triple valve springs.

Double springs with flat wound damper fitted.

the main spring. It also provides very limited assistance in the form of operating pressure, though this is not its intended purpose. Triple springs take matters one step further.

Some engines have valve springs with a flat profile coil fitted inside. These dampers do not act as a supplementary spring. Fitting tightly they merely rub against the inside of the working spring, using friction to control its behaviour. This is a rather poor means of solving valve spring problems: the friction generates heat and wear, which cannot be good.

Another method is to have a spring that is wound with more coils at one end. Or springs which are wound in such a way that the whole spring is tapered; the diameter at the bottom is larger than that at the top. As you have probably gathered, there are myriad options and combinations of valve springs available. This is obviously a specialist field, so heed the word of your cam supplier!

Using stiff springs does cause problems. The stiffer the spring the more force the cam has to exert to open it. This force resists the engine's rotation. Multiply that force by the number of valve springs on the engine and it can add up to a considerable amount of power absorbed; frictional losses, remember!

Stiff springs also put increased load on the cam and follower, causing them to wear more rapidly. The entire valve train is also stressed, causing pushrods, rocker posts, shafts and/or studs to flex and move around. Hardly conducive to accurate valve timing!

INSTALLED HEIGHT & COIL BIND

In order to control the valve train motion as intended, the valve springs must be fitted correctly, preloaded (seat pressure) so that, when compressed by the valve train motion, they are able to offer sufficient resistance ("open" or "over the nose" pressure) to prevent any of the unpleasantness mentioned previously. If the cam has been changed, or high ratio rockers fitted then, in all probability, the springs will also have to be changed, as there may be insufficient clearance between the coils with the new valve lift, or insufficient strength to counter the new accelerations.

When a spring reaches its fully compressed state (ie. all the clearance between the coils is used up) the spring is said to be coil bound.

Alternatively a spring that has been fitted "too loose" (ie. insufficient preload or seat pressure) also cannot function correctly.

Caution! - The spring installed height **must** be checked during assembly of the head and is an **essential** step to avoid possible damage or even catastrophic failure. The installed height is the distance from the spring seat on the head to the underside of the cap: it has an effect on the correct operation of the spring as well as coil bind.

The recommended installed height is specified by the spring supplier. An incorrect height can be remedied by machining the spring seat on the head (increase the installed height) or by adding shims under the spring (reduce the height). Be aware, though, that most alloy heads have thin steel shims fitted beneath the springs as standard. These are not true shims, instead they act as seats for the springs to prevent damage to the softer material of the head.

To check for coil bind you need

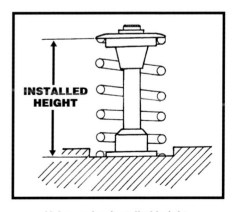

Valve spring installed height.

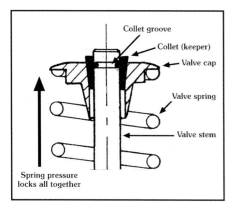

Collet groove
Collet (keeper)
Valve cap
Valve spring
Valve stem

Spring pressure locks all together

Due to the way they interact, valve components must be spotlessly clean before assembly.

to measure the valve spring fully compressed (**Caution!** - be **very** careful!) and subtract this from the installed height measurement (including any shims used). The difference should be a minimum 50 thou' (1.27mm) greater than the maximum valve lift of the cam/rockers. If the spring binds before full lift is reached, valve train longevity will be severely compromised.

The best springs to use for a given cam design are those recommended by the camshaft manufacturer or supplier. The manufacturer will have crunched the numbers and performed the tests to arrive at the optimum spring type for their particular product. The spring recommendations are usually supplied with the cam, or are listed in the camshaft manufacturer's catalogue. Most camshaft manufacturers have telephone helplines available, so consult them if you are uncertain. It's best to heed their advice and, by doing so, avoid later problems.

As an aside, the ultimate solution to the problems of high speed valve spring operation is to eliminate the springs, as has been done in the case of some high revving (currently some 16-17,000rpm!) Formula 1 engines.

The system was invented by Renault Sport, and subsequently adopted by most of their competitors. These engines are of multiple overhead cam design but they have done away with valve springs and instead use "sealed chambers" in the space beneath the cam buckets. An inert gas under a "base pressure" fills this chamber, to hold the valve on the seat (preload or seat pressure). The bucket, in conjunction with its sliding seal, acts as a piston, compressing the gas when the valve is opened. As the gas becomes more and more compressed, it behaves in a similar fashion to a rising rate spring, effectively controlling the valve's motion. Losses from the "base pressure" within the chambers are replenished from a small reservoir carried on the car, through a supply network. Friction is reduced, coil spring problems are eliminated, the valve train mass is reduced and ballistic rpms ensue!

VALVE TRAIN LIGHTENING

Another means of reducing or combating valve train problems is to lighten the moving components of the valve train. By reducing the mass, and hence reducing the forces to be controlled, the spring's job is made easier. Alternative valve and spring cap materials or designs can all reduce

Double springs and a variety of lightweight alloy valvecaps.

A few of the myriad types of valve oil seal - O rings (bottom right); umbrella (top left); top hat (middle).

weight. **Caution!** - Be careful about the use of the aluminium spring caps that are available; with some types, over time, the valve can pull through the cap if stiff springs are used, dropping the valve into the engine. Such caps are best reserved for off-road and track use, where they can be checked and replaced more frequently than would be the case with a road engine.

OIL SEALS

The role of the oil seal is important in that it controls the amount of oil passed into the valve guides. Not an easy task considering they may be sitting in a bath of oil, subjected to all that oil spray flying around. Only a small amount of oil is necessary to lubricate the valve stem. In the case of the inlets, oil control is made more difficult by being subjected to vacuum conditions whenever the valve is open. This can lead to extra oil being drawn through, contaminating the incoming fresh mixture and carboning up the valves and chambers. In cases of worn seals, the engine emits a characteristic blue smoke on starting, or a slight puff during hard acceleration.

For most situations the best oil seals are the flexible rubber variety - be they small 0-rings or "top hat" seals - whichever type is fitted as original equipment by manufacturers. They are more durable, withstanding heat and movement better than the more rigid plastic variety. The rigid types are OK for engines that are frequently rebuilt and have them replaced, but they are best avoided for long term use. Movement caused by worn valves or guides can wear and distort the inflexible seal, allowing past excessive quantities of oil.

With some rubber seals the bodies are rigid, or else utilise a metal clip to ensure correct attachment to the valve guide and provide more positive location. As long as the region that seals around the valve stem is flexible there's no problem.

When a head has been converted to run lead free fuel and has bronze exhaust valve guides, we usually omit the seals, fitting them to the inlets only. The bronze guides run tighter clearances than the cast iron type, so oil control is better. Any extra oil down the guide helps lubrication before being burnt and expelled with the hot exhaust gas, so no contamination of the inlet charge occurs.

The only checks necessary are to ensure that the seal fits snugly around the top of the guide when fitted, or is correctly located - in the case of the O-ring seals. Finally, check that the valve cap is not going to hit the seal at maximum valve lift - an important point if aftermarket rockers or high lift cams have been fitted. The remedy would be to machine the caps (if possible) for more clearance, or fit aftermarket replacement caps if available.

ROCKER ARMS & ROCKER SHAFT

Any wear on the rocker shaft or slop in the rocker arms is going to upset the timing of the valve events and lose power: using an old shaft on a newly assembled engine is silly. The shaft should be dismantled and checked for wear ridges at the rocker positions (usually on the underside of the shaft) and where the posts clamp the shaft (the shaft actually frets inside the posts). Use a micrometer for absolute accuracy, or, at the very least, run your fingers along its length. Any wear at all means the shaft should be replaced (complete, reconditioned rocker assemblies are not expensive), just ensure the correct oil feed is specified, in-line or offset, to suit the head.

You can recondition the shafts yourself by using an oversized replacement, but the rocker arms and the posts will need accurately reaming to suit the larger diameter.

Any wear on the pads that contact the valves should also be removed, especially if you are changing the cam to a performance profile but retaining the original rockers. These wear ridges can cause side loading on the valve and accelerate wear.

Anyone with aluminium rocker shaft posts should change to the stronger and more rigid steel type before any engine uprating is considered.

To blueprint the valve train the head must be fully assembled and fitted to the engine, and the valve (tappet) clearances set. By slowly turning the engine in its direction of rotation, check the middle of each rocker arm pad is centralised on its valve tip at half maximum valve lift. If the head has been skimmed the pushrods may have to be shortened (machine the round pads at the bottom of the pushrods, keeping the same shape as original) to achieve this. If the head has not been too heavily skimmed you may be able to achieve the same result by putting several thin shims (as found under the end pillars of later heads) under the rocker pillars.

The rocker pads must also be centralised over each valve, by putting thin shims between the pillar and rocker or removing material from the edge of the rocker.

Once the rockers are correctly centred, all valve lifts must be checked and equalised. Try to match all to the one with the highest lift (or to the correct specified lift if class rules apply) The pads on the ends of each of the rockers may need to be slightly reshaped, so all the valve lifts are nearly equal (within 0.005in/0.127mm is close enough).

Geometry problems can also arise with OHC engines using finger followers. The cam must be centralised over the pad or lift can be lost and wear accelerated. The relevant manual details correct setting-up procedures.

You can see it takes an incredible amount of time to check and adjust everything until it is within specification, and it would be a pointless exercise with a worn rocker assembly.

As you have probably gathered by now, a great deal of time and effort is needed if you want to be a winner in the horsepower game!

HIGH RATIO ROCKERS

Special high ratio rockers are available for many engines and they give a change in valve train lever ratio. Most are made from aluminium alloy and have needle roller bearings for the pivot on the shaft, and a roller tip to operate the valve. The only problem we see is that the needle rollers are a dubious improvement due to the movement of the rocker being reciprocating rather than rotary. The mechanical loads are therefore taken by a couple of needle rollers on the

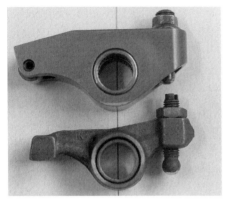

Hi-ratio 1.625 alloy roller rocker compared to standard cast 1.42:1 ratio rocker.

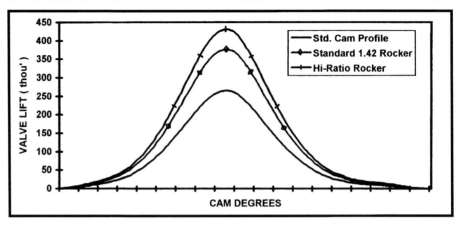

Standard versus high-ratio rockers.

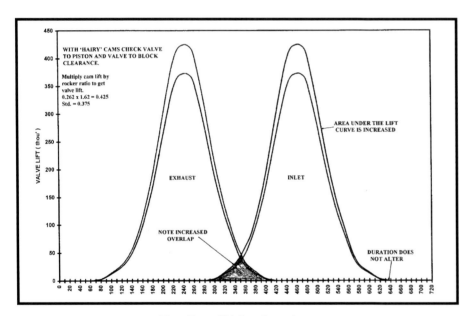

The effect of high-ratio rockers.

underside of the rocker, and are consequently greater due to their small contact area. The rockers become sloppy with use, as the rollers and shaft wear quite quickly.

They don't alter the length of time the valve is open or closed, but do increase the net lift and acceleration of a cam, ie. at any time the valve is lifted just a little bit further with high ratio rockers. The valve springs must be checked for coil bind at full lift (**Caution!** - a minimum 0.010in/

0.254mm clearance between **each** coil of a spring is advisable) and changed if necessary. You also need to ensure that the valves will not hit the block or the pistons at full lift, so bore or piston crown cutouts (valve reliefs) are necessary if not there already. The tappet clearances will also need increasing (multiply by the percentage of change in rocker ratio, ie. 1.42 to 1.625 is 14 % so clearance **must** be increased by 14%)

With a standard cam, fitting high

ratio rockers gives an immediate power gain throughout the rev range. **Caution!** - Consult the camshaft and spring manufacturers about the use of high ratio rockers with high performance camshafts.

COLLETS (VALVE KEEPERS)

The valve collets fit into the groove(s) machined near the tip of the valve, their taper design working with the spring pressure to wedge them into the valve spring cap. The size, number of grooves and angle of the taper of the collets varies with each engine type. Some designs (such as the triple collet groove collets used by Ford) have the collets sized to butt together when installed, allowing the valve to rotate in service. This is to equalise valve wear but can lead to problems for high rpm and race use. The slack fit of the collets allows sufficient movement for the end of the valve to become damaged. It manifests itself as a bulging of the stem above the grooves due to the hammering action experienced, the valves will not pass (easily) through the guides upon disassembly. The solution is to alter the collets so they no longer butt together. Losing the wear reducing rotational facility is not a problem for frequently rebuilt (competition) engines.

Caution! - When assembling the head, ensure the caps, collets and the grooves in the valves are spotlessly clean. Otherwise their clamping, and hence retention, capability will be severely compromised.

Various valve caps, collets (keepers), shims and spring seats for protecting alloy heads.

A few valve nasties! The lightweight headless special, a cracked and burnt exhaust valve and a classic engineering failure due to a poor design.

Rocker arm failure due to over-revving.

Chapter 10
Camshafts

The camshaft's job is to open and close the valves at the right time, in order to fill the cylinders with the fresh mixture before combustion and empty them after combustion. This sounds fairly straightforward, but how the job is accomplished has a significant effect on torque, horsepower, the operating range of the engine and its driveability. A brief description of how the cam works in relation to the four stroke cycle will not go amiss.

The camshaft provides a means of converting rotary motion into the reciprocating motion that opens and closes the valves. The cam is driven off the crankshaft by the timing chain or cam belt, and always rotates at half the crank speed (rpm). It takes two full rotations of the crank, which is one rotation of the cam, to complete the four stroke cycle. It is very important that the camshaft is installed in proper relationship to the crank so the valves are opened and closed at the correct time during the piston's stroke. Setting this relationship of cam timing to crank timing is what is meant by the term "timing," or "degreeing," the cam.

Let's make a few generalised observations, beginning with the exhaust stroke. With the piston on its way down the cylinder during the power stroke, the exhaust valve opens before the piston reaches Bottom Dead Centre (BDC), allowing the remaining combustion pressure to begin leaving the cylinder (called "blowdown"). The piston moves back up the cylinder, pushing the remaining exhaust gas out. Meanwhile, as the piston approaches the top of its travel - Top Dead Centre (TDC) - the inlet valve has started to open before the exhaust valve has closed; this early opening is intended to take advantage of any vacuum that is created in the cylinder by the rapidly departing exhaust gas in order to help start drawing in the fresh air/fuel mixture. The exhaust valve closes fully just after the piston begins its movement back down the cylinder, continuing the induction stroke. If that wasn't enough, the inlet valve then remains open after the piston passes BDC and begins the compression stroke. This takes advantage of the reluctance of the now rapidly moving incoming air/fuel mix to stop, so it keeps piling into the cylinder past the inlet valve, helping maximise cylinder filling. With both valves shut, the piston's motion up the cylinder continues the compression stroke, before combustion is initiated and the whole process begins over again.

With respect to camshafts, compromises are needed to achieve the optimum trade-off between low-speed torque and high-rpm horsepower. During the process of choosing a cam, lift and duration are the most commonly used criteria for determining a camshaft's suitability for a particular application. But, nowadays, more attention is being paid to a cam's "lobe separation angle" - also called the "lobe displacement angle." This has generally replaced the previously used term, "overlap."

CAMSHAFT TERMINOLOGY

The "camshaft lobes" are the lumps (technically called "eccentrics") on the camshaft that are responsible for converting the rotation of the cam into the up and down movement of the valves, via the cam followers, pushrods and rocker arms or, more directly, by means of the followers or buckets with OHC engines. The actual shape ground onto the lobe is termed the "profile," and, even within precisely the same lift, opening and closing points, there are hundreds of different profiles possible.

"The base circle." This is the round portion of cam lobe where there is no lift, and is the area used when tappet adjustments are made. While the follower is on this part of the cam lobe, the valve is allowed to remain on its seat, sealing the port from the cylinder and also transferring some of its heat to the cylinder head via the seat. The tappet clearance allows the

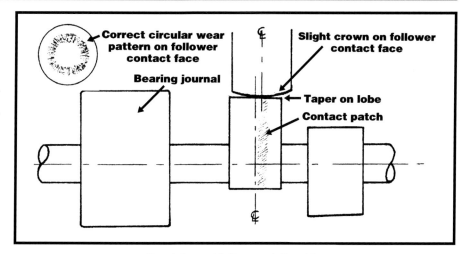

Cam lobe and follower relationship.

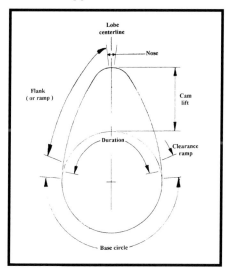

Camshaft terminology.

necessary room for expansion of the valve and valve train through heat.

The "clearance ramp" or "opening ramp" is a short section that takes up the initial clearances between the cam and follower (as well as between the pushrod, rocker arm and valve), providing the transition from base circle to the point on the profile where measurable valve lift begins. Its purpose can be described simply as taking up the slack, as the rate of lift in this region is fairly slow as the shape takes up the various clearances gently, ready to begin lifting the valve off its seat. If a valve train has been set up incorrectly, giving a valve a clearance (lash) sufficient to miss the clearance ramp, then the valve will hammer open and shut too violently, damaging the valve and valve train - even snapping off the valve head!

The region of the lobe that lifts the valve is the "flank" (or "ramp"). This generates the positive acceleration of the follower to produce the lift, followed by a region of controlled negative acceleration (slowing down) until the valve train pauses momentarily at the cam's nose (full lift). The opposite flank then controls the positive acceleration of

the valve train away from full lift, with a further region of negative acceleration prior to lowering the valve gently back onto its seat before reaching the base circle again. Most lobes are ground with mirror image lift curves on the opening and closing side ("symmetrical cam").

An "asymmetric lobe" is one that has been ground with a different lift curve on the opening and closing sides, which means the valve may be opened quite rapidly and then closed more slowly, or vice versa.

"Lift" is the maximum distance that the valve is raised off its seat, expressed in either thousandths of an inch or millimetres. "Theoretical lift" (as used in most cam catalogues) is calculated by multiplying the camshaft's given lobe lift by the (note!) theoretical rocker arm ratio. True valve lift (or "net lift") is the actual physical lift of the valve measured at the valve (usually during blueprinting of the valve train assembly). There is always quite a difference between the two. True lift takes into account the tappet clearances, any rocker arm ratio differences and different rocker pad shapes, as well as pushrod and other mechanical deflections in the valve train. Sometimes the cam lobe lift itself is different, for some reason, to the specification given on paper (a dodgy grind?)

Each lobe of the cam is ground with a slight taper across it. This taper, together with the slight crown on the contact face of the follower, compensates for any misalignment of the follower bores and also spins the follower in its bore to reduce wear between it and the cam, helping to prolong component life.

"Rate of lift" is how fast the valve is opened, ie. inches or millimetres of lift per degree of crankshaft rotation. The diameter of the cam follower base dictates the maximum rate of lift from a cam profile. While lifting the follower, the flank of the cam wipes across the follower's width, from its middle to near to the edge and then back to the middle. Too high a rate of lift can cause the flank of the cam to dig into the edge of the follower, which is not good for long valve train life.

"Duration" is the amount of time the valve is held open, measured in degrees of crankshaft rotation. As advertised, this value varies tremendously, depending on where the cam manufacturer decides to take the start of valve lift.

"Lobe separation angle" ("LSA," alternatively called "lobe centreline angle") is the angle between the centrelines of the inlet and exhaust lobes, expressed in camshaft degrees. This is fixed when the cam is ground, and is gradually becoming more widely used in place of the previously used less accurate term "overlap."

"Overlap" is the period of time when the inlet and exhaust valves are open simultaneously. As mentioned previously, this happens around the end of the exhaust stroke and the beginning of the induction stroke - as

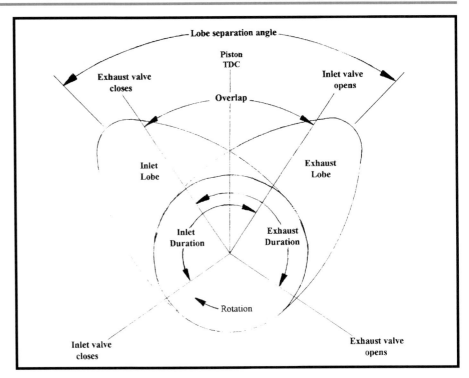

Inlet and exhaust lobe relationship.

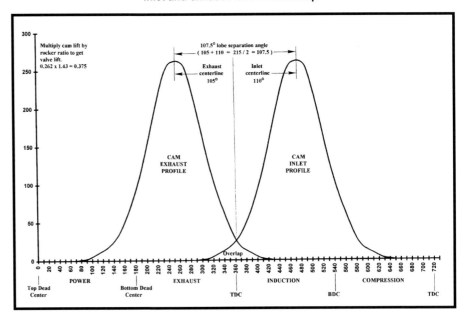

Illustration of "lobe separation angle."

the piston approaches and leaves TDC.

"Single pattern" cams have an identical lobe profile for both the inlet and exhaust lobes. A "dual pattern" cam has different lobe profiles for the inlet and exhaust lobes, usually to overcome some deficiency in the performance of a particular engine's induction or, more commonly, exhaust system. "Scatter pattern" cams use different full lift timing and lobe separation angles for each cylinder of engines with siamesed (joined) inlet and exhaust ports. These have come about in an effort to overcome the flow interference problems siamesed ports present. Whether this has any benefit for conventional engines, we don't know yet, but going to this extent with the cam is probably best saved for race applications of siamese port engines, where every last horsepower is wanted.

"Timing," or "degreeing" the cam. Timing the cam does not change its duration, lift or lobe separation angle, these are all fixed when the cam is ground. It merely allows the point at which the valves are opened, relative to the piston's stroke, to be varied. Moving the cam (in relation to the crankshaft) alters the rpm at which peak torque and horsepower occur. Retarding the cam (opening the valves later in the cycle) moves peak power to higher rpm and can increase horsepower - at the expense of low rpm torque. Advancing the cam (opening the valves sooner in the cycle) has the opposite effect, moving power to a lower rpm and increasing low rpm torque. These effects are not guaranteed, however; they vary depending on the particular engine and camshaft used, so experimentation is the only way to find out.

CAMSHAFT CHOICE

Don't think that a change of cam is essential just because the head has been modified. The standard cam is an often overlooked item that is more than capable of delivering the goods with a strong, tractable, well-mannered road engine. Think about it ...

Giving specifics about cam choice and performance here is impossible, and generalisations about lift and duration are meaningless; the effects differ from engine to engine. For the most accurate advice about camshaft selection, talk to the manufacturers about your application and the type of engine characteristics you require.

Note that advertised duration can include lobe clearance ramps where no valve lift at all is taking place, or given as the duration at 0.050in (1.27mm) cam lift - as used by American cam companies - with many permutations in between. All the different methods of measurement that have been adopted, coupled with the hundreds of different profiles and rates of acceleration available within a lobe shape, means this is a fairly pointless method of trying to compare cam specifications. This is where lobe centreline angles come in (calculations will be found in the appendices). The lobe separation angle can remain constant when the duration or rate of lift (profile) is altered, while the overlap doesn't. So, in effect, two cams with the same lobe separation angle can have dramatically different overlap, depending on their duration.

For a given profile, as lobe separation is widened (say, 107 degrees to 112) overlap decreases, giving increased cylinder pressures and more torque. This gives a smooth tickover (idle) and improved low to mid-range torque, but limits high rpm power. As the lobe separation is tightened (say, 110 degrees to 106) the overlap increases, cylinder pressures at low rpms are reduced, so torque and horsepower are reduced. This gives rise to a lumpy tickover, but better mid-range torque and high rpm power.

If you intend to use high ratio rockers with anything other than the standard cam, look for a profile with between one and three degrees more LSA than standard (from, say, 108 to around 110 degrees). This will counter the effect the higher rocker ratio has of closing up the apparent LSA, due to the increased valve open area (more overlap at TDC). Otherwise, the engine will become more "cammy" than normal.

BEDDING THE CAM

Caution! - The first moments of operation of a new cam are vital in determining whether it will live long, or not. To help cams survive we coat the cam lobes and bottom of the followers with a graphite based high pressure lubricant (eg. Graphogen), but a molybdenum assembly grease is equally good.

Try to ensure the engine starts fairly quickly (set ignition timing, etc, carefully) and, once running, immediately take it to around 2500rpm and keep it there for about 15 to 20 minutes. This high rpm increases the oil supply, reduces the loads on the cam, and gives the lobes and followers time to bed in. We usually rev the engine gently between 2000 and 2700 during this period, to vary the speed and load; do not let the engine idle at any time. If you need to do something to the car, switch off, adjust, then back to 2500rpm for the remainder of the break-in period.

Chapter 11
Fuel & Fuel Supply

On the induction stroke of the engine the fuel that is supplied to mix with the air can be delivered using a carburettor or fuel injection system.

The carburettor is designed in such a way as to use the amount of air drawn in on the induction stroke to meter the fuel, with the engine governed by the throttle butterfly situated downstream of the fuel delivery point. Consideration must be given, during design of the inlet manifold plenum and runners, to the "wet flow" conditions that prevail - namely transportation of the large fuel droplets suspended in the air. These droplets are heavier than the surrounding air, which makes it awkward for them to negotiate turns when travelling at speed. They tend to want to go straight on and can smack into the passage walls. They can also drop to the floor of the plenum or runners if the surrounding air speed is insufficient to keep them in suspension. These inherent design difficulties can lead to unequal fuel distribution amongst the cylinders, with corresponding variations from the intended ideal. The inlet manifold is usually heated to help evaporate the fuel droplets, either by a water jacket using the engine's coolant, or by a "hot spot" heated by high temperature exhaust gases.

An increasingly common alternative to the carburettor is fuel injection. As manufacturers strive to

Or in this case a set of stacks on a mechanical fuel injection system.

Serious horsepower requires serious quantities of fuel, delivered here by a brace of Holley carbs. Note straight manifold runners.

limit engine emissions, whilst maintaining and enhancing engine performance, injection offers a means of eliminating inlet manifold wet flow and carburettor fuel delivery problems. With multipoint injection the fuel for each cylinder is metered directly into each inlet port, ensuring even distribution. Because it is under pressure the fuel is delivered in the form of a fine spray, which mixes far more effectively with the incoming air.

The manifold (or plenum) design has only the movement and supply of "dry" air to contend with. More equal runner dimensions are possible for all cylinders and therefore more use can be made of ram and pulse tuning effects. The throttle butterfly is usually situated well upstream in the system, while the quantity of fuel metered to the air is controlled either through mechanical or electronic means.

Throttle body injection uses a device similar to a carburettor, where the fuel is supplied ahead of the throttle butterfly, so similar wet flow conditions apply to the inlet manifold as would for a carburettor. It is intended as a cheaper means of applying electronic control to the fuel supply.

CARBURATION BASICS

The piston moving down the bore creates a depression (vacuum) in the cylinder and, when the inlet valve is open, in the inlet manifold. As nature abhors a vacuum, the air outside, being at a higher (atmospheric) pressure than pertains in the inlet tract, moves in to fill the depression. On its way in it passes through the carburettor where fuel is added, enabling the engine to produce power.

The main objective of any carburettor is to supply a mixture of fuel and air to the engine in such a form that it can be burnt rapidly and completely. To achieve this the air and fuel mix needs to be supplied in vapour form. Therefore, the carburettor has to be able to break up the liquid fuel, or atomise it, as well as disperse it effectively into the air passing into the engine. How well the carburettor does this has a considerable affect on the engine's combustion efficiency, and hence how well it performs. To top it all off, the carburettor must also be able to supply varying amounts of fuel and air to cope with the engine's varying speeds, and different amounts of fuel to the air depending on the engine's ever changing power requirements.

For complete combustion, the theoretically ideal amount of air to fuel, is called the "stoichiometric ratio" - 14.7 parts of air to 1 part of fuel (by weight). But this does not give either maximum power or minimum fuel consumption from an engine. Maximum power is usually generated with an air/fuel ratio of 12.5:1 (less air) and maximum economy around 16:1 (more air), while a comfortable cruising condition for the engine is around 13.4:1. More modern engines use different A/F ratios to these ideals as manufacturers strive for more power, more economy and less emissions. However, these figures are valid for most 'older' cars.

We can determine the amounts of fuel being delivered to the engine by using an exhaust gas analyser to measure the amount of carbon monoxide (CO) being produced. CO is a byproduct of the combustion process, and there is a known relationship between the amount of CO present in the exhaust gas and the air/fuel ratio. Combining an analyser with a rolling road to simulate driving conditions, allows the mixture supplied by the carburettor(s) to be checked and adjusted, if necessary, to ensure that it is supplying the optimum amount of fuel for best engine performance. Best power is with around a 5% CO reading, best economy with around 0.5 to 1% CO and, for cruising, around 2 to 3% CO is a happy medium.

Too rich a mixture will cause carbon to build up on the combustion chamber and piston crown, as well as interfering with the spark plugs' performance. The excess fuel also contaminates the engine oil, impairing its performance and causing premature cylinder bore wear by washing off the thin film of lubricating oil between the bore and the piston.

On the other hand, too lean a mixture can cause the engine to overheat and may result in burnt valves and damage to the tops of the pistons. A lean mix is also more difficult to set alight, so highlighting any inadequacies in the ignition system - as well as losing engine performance.

An uprated fuel supply system, comprising pump, filter and pressure regulator, will be needed if the engine is going to be making much more power than standard - more detail later in the chapter.

A good quality throttle linkage is also a must for ease of use and setting up, together with a low restriction air filter - again, more information later in the chapter.

We'll cover the two main types of carburettor most commonly used. Firstly, the constant depression variety, typified by the SU but also built by other manufacturers; secondly, variable depression, or fixed jet, variety represented here by Weber or Dellorto carburettors.

SU, OR CONSTANT DEPRESSION TYPE, CARBURETTORS

The SU, used here as an example of constant depression type carburettors, has to be one of the simplest yet most effective carburettors fitted to production vehicles. Its superb design has resulted in a long lasting, robust and mostly trouble-free means of fuelling an engine. But, because of commonplace usage, it has never had the high performance image it deserves. Easily capable of providing power outputs that are only very marginally surpassed at very high rpm by competition type sidedraught installations. Any minor problems encountered with SUs are usually easily rectified by a little straightforward maintenance.

SU carburettors work on the constant vacuum principle (also called "constant depression"). Inside them a sliding piston completely blocks the air passage when the carburettor is not working, resting at the bottom of its travel by the influence of gravity and help from a light spring. The chamber the piston works in is effectively airtight, air only entering or leaving through a hole in the base of the piston which is positioned on the downstream (vacuum) side.

On starting the engine, as the starter turns it over, the carburettor piston's obstruction of the airflow into the engine creates a vacuum on the downstream side. This vacuum is transferred to the piston chamber through the hole. As the pressure underneath the piston is around atmospheric, the pressure difference above and below the piston lifts it from its rest position. The amount the piston is lifted depends upon the amount of airflow into the engine, giving what is, in effect, a self-adjusting carburettor.

A tapered needle is attached to the piston, which fits into a single circular jet beneath it. The fuel is issued from the space between the needle and the jet walls. Since the designers can calculate amount of carburettor piston lift according to the amount of airflow, the diameter and the taper of the needle can be chosen in order to block a certain area of the jet so that only a fixed amount of fuel is allowed through, giving the correct mixture at all engine speeds.

At tickover there is enough suction from the engine to pull fuel from the jet, so no idle circuit is needed. There is also no separate acceleration circuit, the carburettor design again allows for the extra fuel necessary for acceleration to be supplied from the single jet. The central rod that guides the piston is a tube, which is filled with oil. There is a loose fitting brass bush mounted on a central spindle that sits in the oil. So when the engine's airflow demand increases suddenly due to acceleration and the piston tries to lift, it is stopped temporarily (damped) by the hydraulic action of the bush in the oil. The temporarily increased vacuum created by the blockage from the piston draws additional fuel from the jet, richening the mixture.

It is interesting to note that a correctly set up SU has more "fuelling sites" than computer controlled fuel injection systems, ie. more ability to specifically alter fuelling to suit varying load and throttle positions, - not bad for a 'prehistoric' design.

Modifications

The normal carburettor throttle butterfly is a brass disc which is attached to a spindle and then connected to the accelerator pedal via a linkage. On some SU carburettors, the butterfly has a spring-loaded valve fitted to it. The idea behind this is the reduction of engine emissions during periods of engine overrun. If the throttle is closed suddenly with the engine at high speed, the enormous and sudden increase in vacuum immediately draws any fuel that has been stuck on the walls of the inlet manifold into the cylinders, causing a short period of extreme richness and high exhaust emissions. The valve cures this by opening during these extremely high vacuum conditions,

SU throttle butterflies; plain and later poppet valve type.

A view through the carburettor showing the restriction caused by the spindle, screws and poppet valve.

reducing the vacuum and allowing the mixture to burn better with the extra air allowed through. There is a noticeable lack of engine braking in cars with carburettors fitted with these valves. The problem with this design is the springs holding the valves shut get tired with age, causing them to begin leaking when the engine is at tickover, giving an uncontrollably high idle speed.

The valves also present a large physical obstruction to airflow so it's a good idea, from a high performance viewpoint, as well as a better idle, to change the butterflies on carburettors fitted with butterfly valves to plain non-valved ones. Failing this, a temporary solution to the idle speed problem is to solder up the valves so they cannot operate.

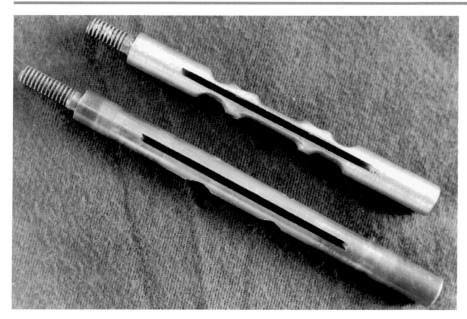

Standard and modified butterfly spindles (modified reduced to 40 thou/1mm aerofoil section between screws).

On the subject of carburettor butterflies, we can mention a few modifications to increase the airflow capability as they come within the range of competent DIY skills. If you take a look at the throttle butterfly in the carburettor, you'll see it is held onto the throttle spindle by two small countersunk screws, which have their ends spilt and spread to stop them falling out. The ends of these screws stick out into the airstream quite a bit, and tests on the flowbench have shown that filing them flush gains a bit of airflow. The screws will then need securing in place with a spot of thread locking compound. Further airflow gains can be achieved by filing the spindle (where it holds the butterfly) evenly on both sides to reduce its width and create a more streamlined shape. The final and most drastic modification involves removing one half of the slotted spindle entirely, leaving a shaft with a flat on which to mount the butterfly. The countersunk retaining screws should be replaced with round head machine screws; again, secure them with threadlock.

It isn't necessary to modify the shape of the SU throttle butterfly itself at all. Thinning it can cause sealing problems if you're not careful, and if taken to extremes can cause it to distort in use.

It can be useful to smooth and deburr the inside of the carb's airflow passage, removing any sharp or protruding edges and blending in any changes of shape, but leave the bridge at the bottom of the carburettor - where the jet is - well alone.

A standard 1.5in SU flows 146cfm on our flowbench, radical modifying of the butterfly spindle gains 10% more flow, while smoothing and blending of the flow passage adds a further 3%, resulting in a flow of around 165cfm. A standard 1.75in SU flows around 204cfm, and responds in a similar positive manner to the same modifications.

The switch to larger 1.75in SUs is only required for very fast large bore road engines with hairy cams or for competition use. The larger air passages lose out slightly in making power in the sub-2500rpm regions, while they only provide a performance gain over the 1.5in SUs at rpms above 6000, when their size allows more airflow.

There are no other modifications that we've tried that increase airflow without actually compromising the operation of the carburettor.

We in no way condone the idea or practice of filing or reshaping the bridge at the bottom of the carburettor where the jet is located in order to increase the flow. It was shaped square edged by the original designers for a reason. They experimented with all the permutations of shapes possible and settled on this flat and square shape, possibly using the deliberately 'tripped' airflow to lift fuel from the jet and help in dispersing the fuel droplets. Reshaping seems a good idea in theory, but if the factory hasn't altered this region in all their redesigns, and as they know best, the message is leave well alone!

If the cylinder head ports have been modified and opened up at the manifold face, the manifold itself will benefit from being opened out to match the port sizes, by removing any obvious step there may be. A carbide burr works best on the aluminium, dipped in - or sprayed with - a little WD40 occasionally to reduce clogging. The same applies if larger SUs are fitted, just remove any step there may be between the carb and manifold. Don't be tempted to block or obstruct the balance passage in the inlet manifold that links the twin carburettors together, it is there to help reduce pressure fluctuations in the manifold caused by the valves opening and closing, and so smooth the carburettors' operation. Flow testing showed a 1cfm gain when the hole was blocked and smoothed over, but the engine lost two horsepower and ran very raggedly on the rolling road.

Cleaning

In most cases, when you look inside the carburettor dashpot you'll see that the inside of the pot is encrusted with burnt oil and varnish from the engine heat (it's brown in colour instead of bright clean aluminium). This must be removed, using choke cleaner or thinners, to allow the piston to slide freely inside the dashpot. Any dirt on the piston itself is also cleaned off at the same time. This accumulation of muck can lose power and reduce performance.

When the dashpots are refitted, refill with engine oil to the top of the rod - low, or no oil, will make the fuel mixture weak during acceleration, causing the engine to cough and splutter for a few seconds.

FIXED JET TYPE CARBURETTORS

The carburettors from manufacturers such as Weber, Holley and Dellorto operate on the same Venturi effect as the SU. But, whereas the motion of the piston in the airflow passage creates a variable (area) venturi with the SU - creating a constant depression (vacuum) at the jet at different engine speeds, these carburettors use a fixed area venturi, creating a variable depression (vacuum) with the varying engine speeds. The venturi is commonly called the "choke." These carburettors can have one, two, or four intakes or barrels. Two and four barrel (V8 engines mostly) carburettors are arranged as primaries and secondaries, controlled by butterfly and linkage arrangement. The primaries and

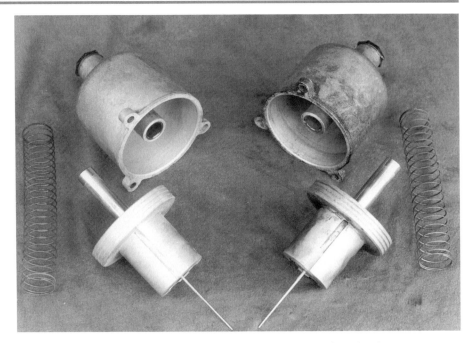

SU carburettor dashpot components before and after cleaning.

secondaries can open simultaneously, or progressively. Four barrel carburettors can have primary and secondary butterflies of the same or different sizes (square or spread bore). The primaries on those with different butterfly sizes, being smaller, are intended for low speed use and economical cruising. The larger secondaries open when rpm is higher or more power is demanded and can be either mechanically or vacuum operated.

Specific carburettor designs operate vertically (downdraught) or horizontally (sidedraught).

A typical twin choke sidedraught has two intake barrels (usually one barrel per cylinder), each containing a venturi or choke (hence the name), that sit either side of a common fuel reservoir (float chamber). The two throttle butterflies are connected by a common shaft to act as one. The air is speeded up in passing through the venturi as it is drawn through the carburettor, the highest velocity being

at the narrowest section of the venturi. The velocity increase reduces the air pressure to below atmospheric. A fuel outlet is inserted into the venturi at its narrowest section, and the low pressure draws the fuel out from the float chamber. By mixing a little air with the fuel (creating an emulsion) before it is drawn out of the jet, the fuel droplet size is reduced and better fuel atomisation results; this effect is achieved by using what are called air correctors and emulsion tubes. The only problem is that, as the airspeed increases, due to increased demand from the engine, the amount of fuel drawn out increases by a disproportional amount and the mixture goes rich. Conversely, when the engine speed is decreased the mixture goes weak. To overcome this, and the other problems of idling, acceleration enrichment and cold starting, the carburettors use a variety of different fuel supply jets and internal air and fuel control passages, in order to deliver correct fuelling at all

Typical twin choke downdraught carburettor.

engine speeds and loads.

With some designs (eg. Weber and Dellorto), chokes can be changed for different sizes, allowing the carb' to be matched to different engine applications. All varieties use interchangeable items such as idle jets, main jets, emulsion tubes etc., meaning they can be calibrated accurately to supply the correct fuelling for the wide variety of engine applications. All of these parts are available off the shelf from a carburettor specialist.

The intake barrels for the sidedraught carbs come in a wide variety of sizes 40, 45, 48, and - Weber only - 50 and 55 (rare!) and these barrel sizes give the carburettors

their name.

All varieties of carburettors are superbly made precision instruments, and they have been around long enough for a great deal of calibration information and knowledge to have been amassed. This information allows experts to supply the necessary calibration parts to suit a particular application, so a carburettor can be ready to install and use straight from the box. However, while the settings will be somewhere near what is required, they will not be tailored to specifically suit your engine. For this fine tuning operation a rolling road session with an experienced operator is a must.

A change of inlet manifold may

be necessary to allow the use of a different carburettor: a large choice is available from factory and aftermarket tuning component suppliers.

With any manifold, new or otherwise, a quick check is necessary before final fitment to ensure that there are no nasty steps where the carb meets the manifold and where the manifold meets the head. These are easily remedied with a little grinding and fettling, as covered previously.

Publisher's note - Another Veloce SpeedPro title *How To Build & Power Tune Weber DCOE & Dellorto DHLA Carburettors* by Des Hammill is available, and will tell you all you need to know about building these carburettors and calibrating them for your specific application.

AIR FILTERS

A considerable restriction to free breathing at the air filter can be encountered before the air ever reaches the carburettor. The standard filters do a good job of silencing intake noise, and paper filters do a good job of keeping dirt out, but the small intakes and filter material seriously restrict airflow. Replacing them with good high flowing filters will produce a performance gain through improved volumetric efficiency.

The choice of filters available falls between the oiled cotton gauze type (eg. K&N, JR filters or Euro Web), or the oiled polyester foam variety (eg. Pipercross, ITG Megaflow, Ramair or Jamex). They all function pretty much on a par with each other and we're not bothered about slight differences in airflow capability: all can easily meet the engine's needs. Choice really comes down to which type best suits your budget or takes your fancy the most!

The cotton gauze filters (K&N solely use cotton) are strong and easily cleaned and re-oiled and, with careful maintenance, will last practically a lifetime. Some of the other similar gauze types may use cotton or a man-made fibre that lasts as long but doesn't clean so easily, but they still flow air and filter very effectively and that's all that really matters. The foam filters can be slightly cheaper but need more careful cleaning and looking after as they are more easily crushed but, again, can last a long time if well looked after. All the varieties from reputable manufacturers are flame retardant, so do not present any fire hazard should the engine ever spit back through the carburettor. Foam filters are not as efficient as cotton gauze of equivalent area, so you will need room to accommodate a larger filter.

Some form of radiused air entry into the carb is recommended: trumpets, stub stacks, ram pipes, whatever. They smooth the passage of air into the carburettor, improving airflow considerably.

Don't even consider running a car on the road without filters. The muck and grit sucked into the engine acts like grinding paste and can reduce engine life expectancy to less than a quarter of normal. Running open unfiltered intakes might be considered clever on the race track, but even mega budget race teams run filters on their engines nowadays to protect what is a large investment, or to avoid losing the race due to wear reducing the power output (yes, it's getting that critical!). Cheap pancake filters with "seat cushion foam" elements do not keep out the dirt, and in most cases restrict airflow as much as the standard factory filters without filtering anywhere near as well! Running with a metal gauze or wire mesh over the intakes is even worse; it doesn't filter out the dirt and reduces the area of the intake, so reducing airflow, and can affect the fuel metering and delivery characteristics of the carbs. So the recommendation is **always** to use air filters and to get them from a reputable manufacturer.

Fitting the high flow filters means the engine does not have to pull so hard against a restriction. That restriction originally caused slightly more fuel to be pulled from the jet, and, as standard, the mixtures were correct for the engine's operating range. Without the restriction to pull against, the mixtures will become slightly weak when the high flow filters are fitted. The mixture will need checking on a rolling road to enable correct adjustments to be made. The change of needles or jets does not mean fuel consumption will increase, the engine's improved breathing actually means that power is produced with less throttle opening than was previously used, so consumption can actually improve. That is as long as you don't enjoy driving the car harder just to hear the gorgeous intake growl from the carbs!

FUEL SUPPLY

When the engine has been modified to produce a lot more power than standard, the original fuel pump will not be able to supply sufficient fuel to keep up with the engine's needs. How to arrive at an estimation of the engine's fuel supply needs is given in the appendices.

We will not mention fuel injection systems here, you will need to consult the specialists or workshop manuals for relevant settings and tuning data.

Good quality pumps from FSE (Facet), Mitsuba, BG Fuel Systems, Holley, and others, are readily

Aftermarket electric fuel pump and fittings.

available, but make sure they use the correct polarity (+ ve. or - ve. earth) to suit your vehicle. They must be mounted near the fuel tank in order to function correctly, as they work properly when pumping the fuel to the carbs from the tank and not when they're having to suck the fuel from the tank over any great distance. Put a filter between the tank and the pump (make sure it offers no restriction to flow) to keep any debris in the tank from damaging the pump.

Make sure the fuel lines are large enough to supply the engine's needs, your pump supplier or carburettor specialist will be able to advise on the necessary sizing.

To complement the uprated fuel pump an adjustable pressure regulator will also be needed. Too much fuel pressure makes it hard for the carburettor float needles to do their job of governing fuel supply, and, if they are overpowered, the carbs can flood. Ideally, a fuel filter should also be included to keep the muck and grit from blocking the jets. Filters and regulators are available as individual items from manufacturers such as Holley, Malpassi or Purolator, or as a combined unit such as the Malpassi Filter King. They can be mounted in any convenient, safe location in the engine compartment.

When using an uprated pump to supply SU carburettors, the fuel pressure needs to be set to 1.5psi/10.34kpa.

When supplying Weber or Dellorto sidedraught carburettors, they require a fuel pressure of around 3.5 to 4 psi/24.13 to 27.58kpa.

You will need a pressure gauge to correctly set the fuel pressure, connect it to the feed pipe for the carburettors and adjust the regulator to the required value with the pump running and the engine off: alternatively, the pressure gauge can be Tee'd into the fuel supply line and used with the engine running. **Warning!** - Watch out for any fuel squirting about when the gauge is disconnected, though.

FUEL FOR THOUGHT ...

Finally, a few extra points are worth covering.

Warning! - Always ensure that flexible fuel supply lines are fuel proof as well as correctly fitted and secure. Flames look all right as paintjobs on custom cars, but not for your pride and joy.

Warning! - Always ensure the drain pipes from SU carburettor (or similar designs) float chamber breathers are fitted, and are routed away from the hot exhaust manifold that often sits right underneath them. Don't have them positioned to empty onto the exhaust pipe, either.

Always check that, at full throttle on the accelerator pedal, the throttle linkage actually does fully open the carburettor butterflies. Try and adjust the cable length to give full throttle with the pedal on the floor, or fit some kind of positive throttle stop so "pressing the pedal to the metal" doesn't overstress the linkage or snap the cable.

If you have no problems, leave the carburettors well alone. Taking them completely to bits for the sake of routine maintenance is not a good idea.

If you are experiencing running problems, ensure the rest of the engine is performing well and is in a reasonable state of tune before meddling with the carburettors, because faults blamed on carburettors are frequently not down to the carburettors at all. A full diagnostic and rolling road session may be worth considering.

If you are going to take the carburettors apart, obtain the relevant manual first. There are many superb specialist books widely available. Each carburettor type usually has umpteen different variations on the same theme, most of which will warrant specialist advice or techniques, which the manual will cover. If in doubt make sketches as you go along to assist reassembly. Have any necessary gaskets ready beforehand, general reconditioning sets are available from carburettor specialists. Clean the outside of the carburettor(s) thoroughly before dismantling, as dirt and grit are the worst enemies of these precision mechanisms. **Caution!** - Carburettors are fairly delicate instruments; excess force may result in stripped threads or distorted components.

A fuel filter in the supply system will help to keep any muck from blocking jets.

Never assume that any carburettor, new or used, is accurately calibrated for your engine as purchased. This is very rarely the case. To achieve optimum performance from the carburettor you need a rolling road dyno session.

When using a non-standard carburettor, a bad throttle linkage can ruin the entire installation, making setting up a nightmare. Superb linkages are available from the aftermarket.

Always use good, high performance, high filtration air filters to eliminate any flow restriction.

Chapter 12
Exhaust System

The exhaust system can be divided into two components, each performing a separate function, each affecting the other.

The first part of the system is the exhaust manifold. Its job is to contain the exhaust gases that leave the engine after each power stroke. These gases leave the exhaust ports as high speed, high pressure, intermittent pulses. In front of each pulse is a high pressure area, greater than atmospheric pressure, and behind each pulse there is a low pressure area, which is lower than atmospheric. The low pressure helps toward drawing out the remaining exhaust gas from the cylinder.

Also created when the exhaust valve cracks open are sudden pulses of energy in the form of pressure waves. The waves travel at the speed of sound, passing through the pulses of exhaust gas The outgoing high pressure waves are initially contained within the individual exhaust pipes.

When pressure waves hit something solid they bounce, or are reflected off it, still remaining a high pressure wave. When the high pressure wave pops out of an open pipe, or suddenly passes from a small area into a larger area, a low pressure wave is reflected back through the pipe. So these waves - high and low pressure - are whizzing back and forth in the pipes regardless of which way the gas is flowing.

The idea is to time the arrival of the returning low pressure to when an exhaust valve is open in order to make use of it. These waves are only generated at the exhaust valve, so the time they take to arrive back at the valve depends on the speed of the wave and how far it has to travel along the pipe before being reflected.

By coupling the pipes from the various cylinders together at a common point (the collector) - which in effect creates a change in area - the low pressure waves created from the individual pipes can reflect back up all

the other connected pipes, and broaden the speed range of the effect.

In theory, the low pressure wave can be used to increase engine efficiency in two ways. Firstly, if an exhaust valve opens to a low pressure area in the exhaust pipe then the flow rate through the exhaust port is higher - a process called scavenging - which reduces the engine's pumping losses. Secondly, during the overlap period when the inlet and exhaust valves are open together (piston near TDC at the end of the exhaust stroke), the low pressure created in the cylinder and exhaust port by the rapidly departing gases will help start the fresh charge flowing through the inlet valve earlier. This increases the cylinder filling, and so increases the engine's volumetric efficiency.

Although the flow of exhaust gas is not steady, the time taken for these pressure waves to reflect back and forth is fairly constant, so the effect will occur over only a very narrow speed (rpm) range. On an engine equipped with a separate exhaust port per cylinder, quite a lot can be done to tailor power delivery using these principles, by playing with the exhaust manifold pipe lengths. The first pipes that make up the manifold are called the "primary pipes" which, ideally, are all the same length. These can join into further individual pipe lengths called the "secondaries" which, in turn, all combine at a single collector just before reaching the exhaust system itself.

Broadly speaking there are three types of exhaust manifold layout. The first uses very long primaries to keep each cylinder's exhaust pulses separate. This pipe length is called the "tuned length" and varies with engine design. The intention is to generate a low pressure signal in each pipe, which arrives back at the exhaust port just as the valve opens for the next exhaust stroke. All the individual pipes come together at a collector, no secondaries are used. This "independent" manifold design promotes mid to high rpm power, and is most commonly used for racing (called a "four into one

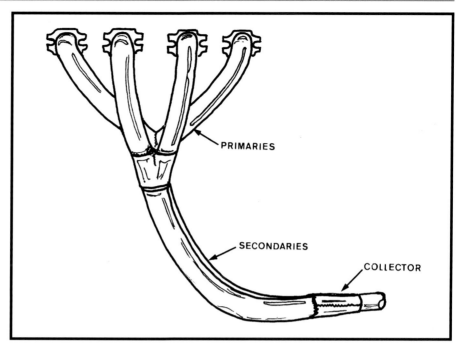

Four into two into one manifold.

manifold" on most four cylinder engines with separate exhaust ports for each cylinder).

The second type of manifold has much shorter length primary pipes which are paired together - usually cylinders one and four and cylinders

two and three - into two secondary pipes (which can be of similar length to the primaries) that finally come together at the collector. The entire manifold length is usually similar to that of the long primary "independent" version. In theory, the exhaust pulse from number one cylinder is helped along by the low pressure area in its primary, generated by the earlier exhaust cycle of number four cylinder, and vice versa. The same effect applies between cylinders two and three. This design is known as an "interference" manifold (or "four into two into one," or "tri-Y" design), and the effect is to enhance low to mid-range power. This design is ideal for a roadgoing engine.

The third type of manifold evolved from the motorcycle world and is intended to combine the best of both worlds. The primaries are very long, as per the independent manifold, but then connect to a pair of secondary pipes - again similar in

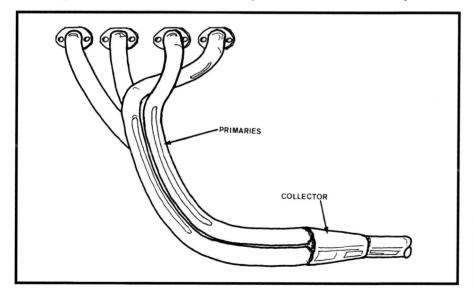

Four into one manifold.

length to the interference manifold (or up to half the length of the primaries) - before the collector leading into the exhaust system. This type of manifold design is still somewhat uncommon in the car world.

The diameter of the manifold pipes can also influence exhaust gas flow characteristics. Small bore pipes, as used on road manifolds, keep the gas velocity high, helping produce a stronger negative pulse in the system and improving cylinder scavenging. Large bore pipes are generally used to allow the greater volume of exhaust gas produced at high rpms to flow without restriction, whilst maintaining a similar pulse assistance effect.

As you have probably gathered, there are many combinations of design, pipe length and pipe diameters to juggle with. The problem with all these tuned lengths and pipe sizes is that they only have an effect over a fixed, rather narrow, rpm band. The specific tuned length varies depending on engine size and the rpm at which the effect is wanted. At another rpm range the effects may be all wrong and the engine can lose power, experiencing flat spots or hesitation whilst accelerating. Throwing in the effects of engine modifications, such as head swaps, cam changes, etc, adds yet more variables for inclusion in the exhaust manifold design equation. Coming up with a design to suit every case would be near impossible, so effective compromises have to be reached. These compromises are usually arrived at after a great deal of exhaustive testing (no pun intended!)

The second component is the actual exhaust system - the piping that runs from the manifold collector to the rear of the car. Again, this has two main functions. One, it enables the fitment of silencers to keep the noise generated by the pressure waves and high speed exhaust gas to an acceptable level. Two, it contains the exhaust gas in a closed system until it can be released to atmosphere from a position which will not gas the vehicle's occupants or present a hazard to others.

Similar effects to those described for the exhaust manifold also occur in the system itself, although they are usually created too far away from the engine to be of any real benefit. It is the positioning of the first silencer that has the most effect on the power curve of the engine in this respect.

Silencing exhaust noise is usually achieved by baffles or absorption. Baffle design silencers usually contain various obstructing plates with holes in, and empty chambers that serve to break up and scatter the pressure waves until they lose their energy. The problem is that in some cases this tortuous and convoluted path can hinder the flow of the gas and create the flow restriction called "backpressure." These type of silencers are more widely favoured in the USA, where silencer designs have largely eliminated this problem, whilst still silencing effectively.

The absorption silencer allows the exhaust gas to flow freely "straight through" it, passing along a perforated tube. The pressure waves are allowed to expand through the perforations and are slowed and diffused by the densely packed, heat resistant wadding that surrounds the tube.

The wadding that makes up the sound absorption layer in the silencer box also plays another important power related role, as well as quietening the exhaust. If the back box becomes hollow - due to deterioration of the wadding over time and its being blown out the tailpipe - the car can lose a lot of power! The exhaust note will also be noticeably louder, though you may not have noticed when driving due to the noise increase being gradual as the silencing effect of the wadding decreases. A quick method of checking is to tap the box with a metal rod/bar or a large spanner, it will obviously sound hollow, rather than the duller sound made by a full box.

The first box removes the low frequency noises from the exhaust. The role of the rear silencer is to filter out all the irritating higher frequencies, the really nasty, crackly, raspy notes.

Most standard cars have cast iron exhaust manifolds, efficient or otherwise. Cast iron is used for its excellent acoustic absorption and therefore noise damping ability (some tubular manifolds "ring" under certain conditions) and for its superior heat transfer characteristics. It is well suited to handling the very hot, chemically very unpleasant mix that is belched out of the engine on every exhaust stroke, and is cheaper for mass production than actually fabricating a complicated manifold shape. Having said that, some modern performance cars have tubular manifolds factory fitted.

As a bit of an aside, we ran a championship winning MGB - in the standard race class - for a whole season using exhaust insulating wrap. The thinking was that the engine bay would run cooler, as the exhaust heat was now contained in the pipes, and so allow the engine to make more power. After fitting the wrap no difference in power was recorded with the car on the rolling road. But we left the wrap on as we thought that under actual race track conditions, with the very high underbonnet temperatures that racing produces, there would be more power available. The actual theory of reducing the intake air temperature is correct, every 3°C reduction gives about 1% more power.

Unfortunately, with hindsight, we went about it the wrong way. After the end of the season when it came time to fit the latest specification engine ready for the next season, the cast iron exhaust manifold would not fit back on the head. It had warped by around 0.25in/ 6.3mm, due to the extremely high temperatures it had encountered. We had created our own problem. On the one hand, the underbonnet temperatures were lower so perhaps giving a little more power. On the other hand we were actually losing power by overheating the exhaust internals (by trapping all that extra heat with the wrap), making the cylinder head hotter, which heated up the inlet ports and then the fresh mixture. So, scrap one twisted exhaust manifold, time for a rethink ... Our simple solution was to fit a thick aluminium plate, as a heat shield, in order to physically separate the inlet and exhaust manifolds. This shield then allowed each component to function as was originally intended.

Further testing on another car fitted with a Ford crossflow engine - where the inlet is on the opposite side of the head to the exhaust so heat doesn't affect it so badly - the car showed a 3bhp at the wheels gain once the wrap was removed. Make of it what you will, but it appears not a good idea to discard the standard heat shield from the car if one was fitted, it was put there for a purpose. Maybe alternatives such as lagging the pipes had been tried and rejected by the original designers. Who knows?

If you're really keen to have a go at making your own manifold and system, then we would recommend investing in one of the specialist books about exhaust system design that are available - it is really quite a complicated subject that goes far beyond what we've covered here. Exhaust components, pipes, bends, silencers, etc, are readily available should you wish to make your own manifold and system. Your best bet would be to use the original as a template and go from there. Even then it will take an awful lot of trial and error, as well as a lot of engine dyno time, to come up with a successful design.

Alternatively, "performance" manifolds and systems are widely available from the aftermarket. For road use, we recommend erring on the conservative side when it comes to manifold pipe diameters. A big bore manifold may seem the choice for more power, but it is more likely to hurt performance than improve it in most road applications.

Chapter 13
Ignition System

The ignition system initiates combustion of the air/fuel mixture in the cylinder. However, the entire charge doesn't burn at once (the "bang" part of the four stroke analogy should read "burn"), so the mixture needs to begin burning just before the piston reaches TDC. That's why ignition timing is described in terms of advance, ie. 14°BTDC.

The reason for the advance curve is because each rpm level has an optimum setting for performance. Assume it takes a certain amount of time for the mixture to burn. Ideally the burn should achieve maximum cylinder pressure just after TDC on the power stroke. At low rpm there is plenty of time for the burn to happen as the piston speed is low, so little spark advance is needed (say 8 BTDC). At high rpm the piston speed is much higher, consequently there is much less time for the burn to happen. To achieve optimum cylinder pressure the spark must initiate the burn earlier (say 34°BTDC). Vacuum advance adds extra ignition timing at part throttle for better fuel economy and performance.

In providing the advance facility by mechanical means (inside the distributor) a slight compromise is required. The optimum performance setting cannot always be achieved for all engine rpm, though the factory is usually pretty close. Obviously, with any major engine changes these optimum settings will also change.

The alternative is to provide the advance facility by electronic means, a microprocessor triggering the ignition based upon inputs from various sensors on the engine. The spark supply can be from a single coil through a distributor, or the increasingly utilised one coil per engine cylinder system.

Fortunately, most engines are fairly insensitive to small changes in ignition requirements and the standard distributor, **in good condition**, can cover most roadgoing applications. Where unsuited, the mechanical components of the advance/retard mechanism can be altered to get nearer to what's needed.

Microprocessor based systems are more effective in meeting the exact ignition requirements of the engine for optimum efficiency at all speeds and loads, especially in terms of meeting the increasingly stringent emissions laws. For race use (where permissible) full electronic control of the ignition can deliver advance/retard characteristics impossible to achieve mechanically, improving driveability and outright power. Such systems are difficult to reprogram effectively, though.

Publisher's note - Another Veloce SpeedPro title *How Build & Power Tune Distributor-type Ignition Systems* by Des Hammill is available. It covers all aspects of ignition systems in high-performance applications and will tell you how to modify for your distributor and how to find the optimal ignition timing and advance for your individual application.

POINTS TYPE DISTRIBUTORS

The points never seem to last very long in distributors, which, coupled with distributor shaft bearing wear, causes the distributor to become inefficient with old age. However, as long as the mechanical and vacuum advance are still working correctly, it is well worth fitting an aftermarket electronic ignition system to do away with the points and they work well in spite of any spindle wear present in the distributor. A wide variety of excellent systems are readily available from, for example, the likes of Lumenition, Accel, Mallory, MSD, Piranha, Micro Dynamics and Aldon.

When it comes to buying a new distributor, most parts and accessory specialists carry a range of "tuned" ones, although these will not be suited specifically to your engine - this can only be done by a specialist with a rolling road or dyno and a distributor machine.

For road use it is best to retain a vacuum advance facility to give good part throttle cruising economy.

For race use with wild camshafts and large carburettor chokes, the engine runs very poorly at low rpm. This situation can be improved by having more advance at low rpm with less mechanical advance in the distributor, so the total remains the same. Here, the vacuum advance can cause slight fluctuations in ignition timing, so it's best to run a distributor without it.

MAPABLE ELECTRONIC IGNITION SYSTEMS

These require specialist computer

equipment and access to a rolling road or engine dyno facility. Here, the engine rpm can be stabilised under various load conditions and the advance for best performance dialled in. Without the correct test and monitoring equipment damage may unwittingly be caused to the engine.

Currently, ignition mapping is best left to specialists, though the rate of simplification of the mapping process may open avenues the home tuner can explore safely in the future.

COILS

If the standard factory coil is of the high output variety it is adequate until it wears out, when it is best replaced with a sports or high performance coil. Be sure the type chosen is suited for use with the engine's triggering system - points or semi-conductor switching. For some early cars, the ballast resistor must be bypassed in order for these coils to function correctly; usually meaning running a new 12V supply direct to the coil from the ignition switch, bypassing the original power feed cable totally.

SPARK PLUGS

The spark plug is the means of getting the ignition energy into the chamber to initiate combustion, whilst withstanding high voltage, temperature, pressure and vibration and operating in a chemically unpleasant environment.

In order to function effectively the plug must be capable of self-cleaning to remove the chemical and soot fouling, otherwise the spark will use this as an alternative means of reaching ground and the mix will not be ignited. By using some of the heat from the engine the deposits are burnt off, the rest of the heat it must

A few of the many spark plug designs currently available.

dissipate effectively or risk the hot spots causing pre-ignition. Different types of engine mean different plugs are needed to allow self-cleaning to occur; what may run hot enough in one engine will not reach the correct temperature in another. This is what is meant by spark plug heat range. A plug for a hot engine means it transfers the unneeded heat away more efficiently in order to maintain its working range - the plug runs cooler. A plug for a cold engine means it must retain more heat - the plug runs hotter.

There are many different types and brands of spark plug available, we always use NGK which seem slightly above average in terms of resistance to fouling, power output and life, but the equivalent types from other manufacturers are also satisfactory. It depends upon personal preference which you choose. For road use, where the car is likely to see mainly town and urban use or predominantly short journeys, use plugs which run hotter and so keep cleaner with this type of motoring. For a car that is used mostly for longer journeys, such as open road countryside and motorway (freeway) driving one grade colder is usually best.

For race use, the grade to fit depends on the engine specification and which plugs produce the best performance. The variety we prefer use a very thin platinum centre electrode that only requires a low voltage to create a spark, functioning well in an engine with high compression pressures. But they are expensive and their life is short - only four or five ten lap races!

All spark plugs usually perform best with the manufacturer's recommended gap.

SPARKPLUG (HT) LEADS (WIRES)

The best type of plug leads are the ones with the long spark plug boot that fits tightly and covers the entire body of the plug. These are better at keeping out moisture and dirt, both of which can create an easier path for the spark to take (down the outside of the ceramic) and not through the plug to do its job as intended.

A little hand cleaner and a cloth works wonders for keeping the leads clean, to reduce the chance of spark leakage and for cosmetic purposes.

The entire ignition system should be kept clean and well maintained. Dirt and moisture provides an alternative, easier route for the spark to take, which reduces firing voltage and current available at the plugs, upsetting engine performance.

Plug lead construction has changed significantly over the years. In the 60s and 70s, the high-performance lead was copper-cored with an externally mounted suppressor (copper has poor suppression qualities). Unfortunately, copper also has poor heat resistance, and the leads can melt ...

To overcome these problems, the carbon core lead came into wider use. It has good suppression properties, but still has poor heat resistance and is subject to internal breakdown after as

little as 18 months of use.

More recently, the silicone lead has come to the fore, offering good performance allied with excellent heat resistance and long life.

PRE-IGNITION & DETONATION

While on the subject of ignition the causes and effects of detonation and pre-ignition are included, as the ignition timing has the greatest affect on the problem.

Under normal circumstances the spark ignites the mixture in the combustion chamber. Burning of the mixture starts with a small kernel of flame which steadily expands and spreads throughout the chamber. As the flame advances across the combustion chamber, the unburned gases ahead of the flame front are subject to increasing temperature and pressure. The last part of the mixture to burn is called the end gas.

"Detonation" is the uncontrolled explosion of the end gas as it ignites spontaneously due to excessive heat and pressure in the combustion chamber, and is characterised by a pinking metallic sound coming from the engine. The noise is caused by the shock wave from the explosion passing through the engine block. The energy released during severe detonation is enough to knock core plugs out of the block, or even punch holes clean through the top of the piston. Light detonation can slowly nibble down the side of the piston, leaving the rings exposed (which burn away due to pre-ignition).

Detonation is usually caused by having too much ignition advance or too high a compression ratio. A weak mixture doesn't help the situation, but

on its own does not usually cause detonation.

"Pre-ignition" is when the mixture in the combustion chamber is ignited from a source other than the spark plug. It is caused by something glowing red hot in the combustion chamber area, such as sharp edges left after skimming the cylinder head or block, or by an overheated exhaust valve. Occasionally, sharp edges left after fitting valve seat inserts can also be the culprit. It is best, therefore, to deburr anything that feels sharp before assembly to preclude this happening. On an old high mileage (or newer engine subject to short journeys with continual cold starts), pre-ignition can be caused by carbon build-up on the piston crown or combustion chamber surfaces. This is exacerbated by lead-free fuel, due to the very hard combustion products deposited in the chamber and on the piston crown. The only solution is to decoke the cylinder head and the top of the piston, but don't remove the carbon from around the edges of the piston - it acts as a compression and oil seal on very worn engines!

Occasionally the extra heat build up caused by detonation can lead to the onset of pre-ignition.

The pre-ignition that most people notice is when the engine runs on when the ignition is switched off. On some cars this can be caused by a high idle rpm, a weak mixture or retarded ignition timing, or even all three together. The problem is most commonly caused by excess heat in the exhaust valve, when the engine is at idle, which is part of the reason for using bronze guides on lead-free conversions in order to take away the excess heat and reduce the likelihood of this happening.

Chapter 14
Head Cases

This section of the book covers modifications we have developed for some popular cylinder heads. Where possible, we have included power figures, flow figures and photographs to highlight the important features of each head. Where given, the power figures are as observed on a Clayton water brake rolling road dynamometer.

FORD CVH - HEMI CHAMBER

The inlet ports are too small in relation to the 42mm valves. The port needs to be opened up to improve airflow. However, if you go too far, you'll find waterways lurking adjacent to the valve guides!

As the head is aluminium, the seat inserts do not blend neatly into the valve throats. Simply fettling the throats and modifying the back of the valves to produce a smooth flow path gives a extra 10bhp at the wheels on an otherwise standard 1.6 injection

Combustion chamber.

engine. Opening the ports and grinding the guide boss flush with the port wall adds another 8bhp at the wheels.

Inlet seat/throat: 1mm (40 thou) 30° top cut blended into chamber

Modified inlet valve throat.

curvature, 1.3mm (52 thou') 45° seat, 1.5mm (60 thou') 60° bottom cut radiused into port throat.

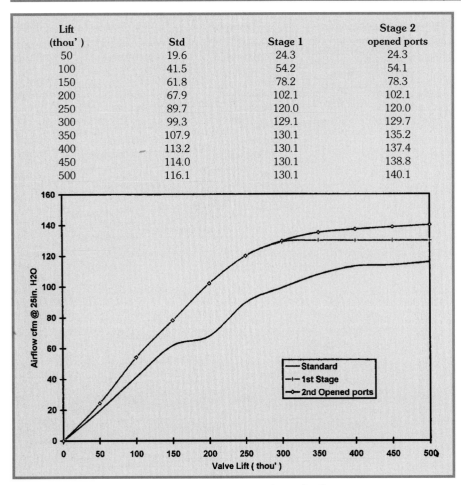

Lift (thou')	Std	Stage 1	Stage 2 opened ports
50	19.6	24.3	24.3
100	41.5	54.2	54.1
150	61.8	78.2	78.3
200	67.9	102.1	102.1
250	89.7	120.0	120.0
300	99.3	129.1	129.7
350	107.9	130.1	135.2
400	113.2	130.1	137.4
450	114.0	130.1	138.8
500	116.1	130.1	140.1

Inlet port flow @ 25in H2O. (Ford CVH).

FORD V6 3-LITRE 'ESSEX'

In standard form these V6 heads are dreadful. The ports are very restrictive and the throats are not compatible with the valve size.

To make the head really flow, the inlet ports must be opened out as much as possible without finding fresh air. The guide boss should be ground flush with the port and a large 43mm (1.69in), or bigger 46mm/1.81in), inlet valve needs to be fitted.

The exhaust valve size is adequate, as is the port. All that is required is 3-angle valve seats and to smooth the guide boss and tidy the port.

With a fast road cam, freeflow air filter and exhaust and fitted with big valve heads, bhp goes up from the standard 110 to around 175 at the wheels.

Inlet seat/throat: Unshroud chamber to gasket line around inlet valve and 10mm elsewhere. 1.5mm (60 thou') 30° cut, 1.5mm (60 thou') 45° seat, 1.5mm (60 Thou') 60° bottom cut radiused into throat.

Standard inlet port - what you see is what you get!

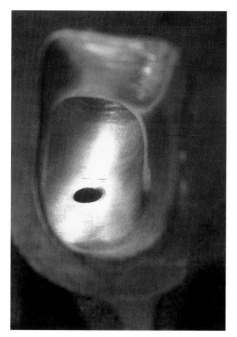

Modified inlet port.

Lift (thou')	Std	large inlet valve.
50	19.3	23.4
100	39.7	46.4
150	60.2	69.1
200	76.6	89.7
250	89.2	108.4
300	94.9	124.0
350	98.7	136.8
400	101.1	143.6
450	103.2	148.7
500	104.0	152.1

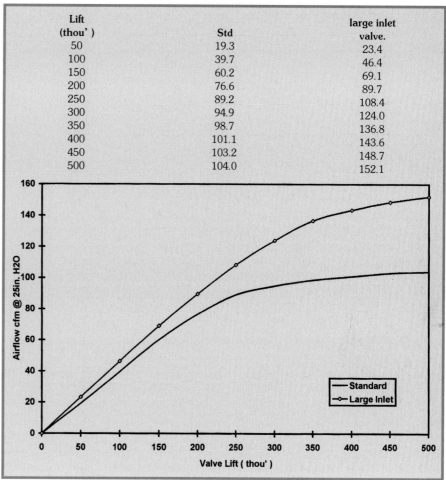

Inlet port flow @ 25in H_2O. (Essex V6).

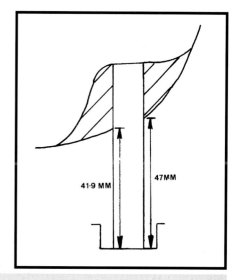

Essex V6 guide boss modification.

FORD 2-LITRE SOHC 'PINTO'

The standard head responds very well to simple or full modification work. In standard form the injection head is nearly as good as a non-injection head with the simple modifications outlined below.

Simple modifications: Cut 3-angle seats and open out the throats to match, smoothing the seats into the throat, especially on the short side of the inlet port. Modify the valves with a cut at 30° from the seat to the valve back, to break the sharp edge. Typical bhp improvement on an otherwise standard (Sierra type exhaust manifold) engine is 18%.

Full modification: To get the best results the head needs large inlet valves, 44.5mm (1.75in). The guide bosses must be ground down flush with the port. The natural angle produced from blending in the increased valve seat diameter down to the guide boss works very well with cleaned up standard shape chambers. Expect a 25% + increase in bhp.

Inlet seats/throat: Stage 1: 2mm (80 thou') 30° top cut, 1.5mm (60 thou') 45° seat, 2mm (80 thou') 60° bottom cut blended into throat.

Stage 2: 1mm (40 thou') 30° top cut, 1.3mm (52 thou') 45° seat, 1.5mm (60 thou') 60° bottom cut blended into throat.

Standard and modified throats and chambers.

Lift (thou')	Std	Injection	Stage 1	Stage 2
50	19.9	23.0	21.4	22.2
100	40.5	40.4	43.5	48.0
150	59.4	62.2	66.3	73.7
200	75.9	83.3	88.4	96.4
250	93.1	105.1	103.9	114.8
300	103.8	119.2	118.0	129.6
350	112.9	125.6	127.9	141.6
400	119.1	129.1	135.7	149.9
450	124.3	133.1	139.7	154.4
500	127.0	135.2	145.4	157.5

Inlet port flow @ 25in H2O. ('Pinto'.)

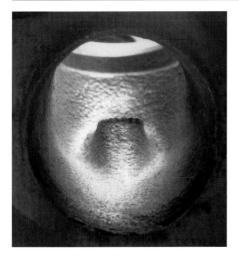

Standard inlet port.

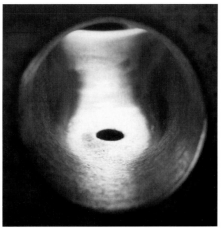

Modified inlet port.

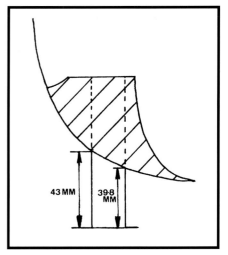

Pinto guide boss modification.

FORD 'KENT' CROSSFLOW 1.6

The inlet port of the 1.6, GT. and Ghia "37 casting" in standard form is slightly curved, aiming the flow towards the bore wall. Flow testing revealed that flow was enhanced by the wall proximity and that extensive work was required to produce effective results when trying to bias the flow direction to compensate: work unnecessary for anything other than outright (race) performance. Standard seats are wide and the casting quality poor in the throat region in particular. Cut 3-angle seats and open and smooth the throats to match. Modify the valves with a 30° back cut. Expect a 15% bhp increase.

Full modification requires considerable material removal to reshape port. Expect a 30% bhp increase.

Inlet port/seat: Seat and throat: 1mm (40 thou') 30° top cut, 1.3mm (52 thou') 45° seat, 1.5mm (60 thou') 60° bottom cut blended into throat. Guide boss retained but port opened up around it, remainder cleaned and smoothed, step at manifold face

Lift (thou')	39.5 mm std	Seat and throat (boss retained)	39.5 mm modified
50	18.4	19.1	21.2
100	37.3	40.0	47.9
150	56.3	61.9	67.4
200	70.9	81.8	86.0
250	82.1	96.8	100.5
300	91.0	110.2	112.6
350	96.7	116.9	120.2
400	99.3	117.1	126.6
450	99.4	114.8	131.0
500	---	---	137.8

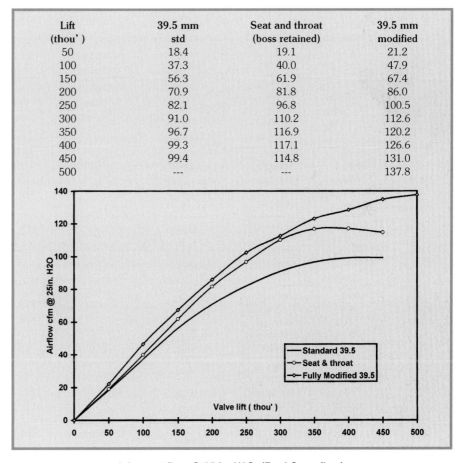

Inlet port flow @ 25 in. H2O. (Ford Crossflow).

removed. Valve backcut.

Fully modified: 1mm (40 thou') 30° top cut, 1.3mm (52 thou') 45°

seat, 1.5mm (60 thou') 60° bottom cut blended into throat. Guide boss removed and port opened as

necessary to create round tube. Valve has 30° cut blended into back to reduce seat width.

VAUXHALL 1.8 8-VALVE

As cast, the 1.8 head offers a lot of potential. With very little effort generous power increases are available. Simply cutting 3-angle seats for inlet and exhaust, blending the throats and backs of the valves increase the bhp at the wheels from 91 to around 108.

Fitting a 43.2mm (1.7in) inlet valve and smoothing the ports, especially around the guide bosses, combined with reducing the guide height, will add another 8bhp at the wheels.

Lift (thou')	Std	Stage 1	43.2 mm Stage 2 large inlet
50	19.6	20.5	20.9
100	38.8	42.1	45.8
150	55.7	66.7	70.8
200	76.5	87.9	93.1
250	93.4	102.4	109.0
300	104.3	115.6	121.8
350	114.5	126.9	131.7
400	121.1	135.8	141.8
450	125.6	140.9	148.1
500	128.4	143.5	153.3

Standard inlet port.

Modified inlet port.

Inlet port flow @ 25 in. H2O. (Vauxhall 1.8 8-valve).

Inlet seats/throat: Std valve: 1mm (40 thou;) 30° top cut, 1.25mm (50 thou') 45° seat, 1mm (40 thou') 60° bottom cut blended into throat.

Large inlet: 1mm (40 thou') 30° top cut, 1.3mm (52 thou') 45° seat, 1.3mm (52 thou') 60° bottom cut blended into throat.

Standard throat, seat and chamber.

Modified throat, seat and chamber (spot the crack).

FIAT TWIN CAM 8-VALVE

In standard form the inlet seat is too wide, making the throat too small. Cutting 3-angle seats and opening up the throats to match works very well. Smooth the ports and grind the guide flush with the port. The inlet valve must remain tulip shaped to be effective.

Expect around a 15bhp increase at the wheels with everything else standard.

On a race engine the modified head would give around a 25% increase in bhp over the standard item.

Inlet seats/throat: 1mm (40 thou') 30° top cut, 1.5mm (60 thou') 45° seat, 1.25mm (50 thou') bottom cut blended into throat.

Lift (thou')	Std	Fully Modified
50	20.2	22.0
100	41.2	44.7
150	64.3	67.4
200	87.9	89.3
250	108.2	112.1
300	119.2	126.3
350	122.8	135.2
400	125.8	145.4
450	128.6	154.2
500	131.6	160.1

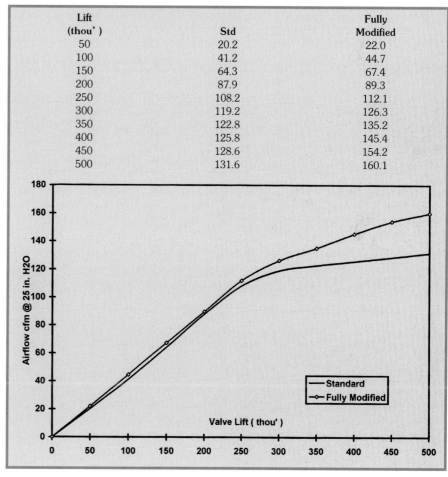

Inlet port flow @ 25 in. H2O. (Fiat twin cam 8-valve).

Standard inlet throat and seat.

Modified inlet throat and seat.

FIAT X1/9 1500

This head responds really well to a full modification programme. With no other changes the power at the wheels increases from 62 to 84bhp.

Chamber modifications: cut the chamber on the plug side of the head to form a straight line from the head face to the original starting point of the chamber wall. On the opposite side lay back the chamber to 0.25 x valve diameter clearance from the inlet valve - **do not** undercut the head gasket !

Port mods: grind the guide and boss flush with the port and generally smooth the ports.

Valve mods: the standard valves are an odd shape, dress the 45° seats down to 1.5mm (60 thou') with a blended 30° cut. 3-angle seats are needed.

Lift (thou')	Std	Fully Modified
50	16.3	17.3
100	33.5	37.7
150	52.1	59.2
200	69.1	80.0
250	80.7	92.3
300	84.9	101.2
350	85.8	105.6
400	86.5	108.5

Inlet port flow @ 25 in. H2O. (Fiat X1/9).

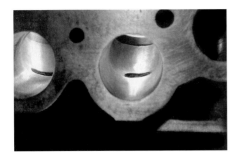

Modified inlet port.

Inlet seats/throat: 1mm (40 thou') 30° top cut, 1mm (40 thou') 45° seat , 1mm (60 thou') 60° bottom cut blended into throat.

TRIUMPH TR7 & DOLOMITE 8-VALVE

In standard form the head looks awful but it flows quite well. To improve it, try to equalise all the port shapes. Grind away the guide boss and use a 7° bulleted valve guide. 3-angle the valve and seat. No chamber mods are required with the standard valve size.

Inlet ports showing poor casting and insert protrusion.

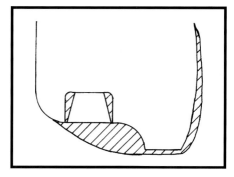

TR7 port and guide boss modifications. (Remove shaded areas).

Lift (thou')	Std.	Modified.
50	16.8	21.0
100	37.4	41.2
150	54.2	64.0
200	74.1	85.2
250	91.4	103.3
300	100.9	115.8
350	109.0	122.4
400	112.5	125.3
450	118.4	129.8
500	119.5	132.0

Inlet port flow @ 25 in. H2O. (Triumph TR7).

These changes will give approximately 15% more bhp (raise the power at the wheels from 85 to 98). This engine responds well to camshaft changes.

Inlet seat /throat: 2mm (80 thou') 30° top cut, 1.25mm (50 thou') 45° seat, 1.25mm (60 thou') 60° bottom cut blended into throat.

TRIUMPH 1500 SPITFIRE/ MIDGET

To make the head work you need to use the large 36.4mm (1.43in) inlet (Spitfire) not the 35mm (1.38in) variety.

Remove the guide boss from the inlet port and use a guide that has been 7° bulleted. The chamber wall on the spark plug side of the head needs to have the chamfer removed to create a straight line from the head face to the original wall starting point in the chamber roof. 3-angle seats and a generous radius on the short side of the port work very well.

Typically, a standard MG Midget 1500 produces 52bhp at the wheels. Add a mild road cam, free flow filters and exhaust and a large inlet valve head and you can expect around 85bhp at the wheels.

Inlet seat/throat: 1mm (40 thou') 30° top cut, 1mm (40 thou') 45° seat, 1mm (40 thou') 60° bottom cut blended into throat.

Lift (thou')	Std 35 mm valve.	Modified 36.4mm valve.
50	15.6	17.2
100	32.2	37.5
150	49.0	55.0
200	64.9	70.0
250	73.5	82.1
300	77.9	89.5
350	82.5	99.4
400	86.7	107.8
450	89.6	114.4
500	90.9	115.3

Inlet port flow @ 25 in. H2O. (Triumph 1500).

Standard and modified inlet ports.

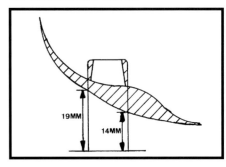

Modified large inlet valve throat.

1275 BMC/BL/ROVER A-SERIES

In standard small valve (33.4mm/1.31in) form the head doesn't work very well, in modified form it works fairly well, with a 35.6mm (1.401in) inlet valve it's brilliant!

Blend the guide bosses into the throat, remove casting differences from the ports and blend any change of direction in the siamese section, ensuring that the port split is symmetrical. For road use, to avoid pinking, clean the casting marks from the chamber. For race use unshroud the inlet valve chamber wall by 0.25D (0.25 times the valve diameter) - or to the gasket line, whichever comes first . Unshroud the exhaust valve by cutting the wall back to 0.2D or the gasket line, again, whichever comes first.

In standard form a 1275 MG Midget makes 45bhp at the wheels. Fit a big valve open chamber head, free flow exhaust, Weber 45DCOE carburettor and a Kent 286 (fast road) cam and you will increase power to 96bhp+ at the wheels.

Inlet seat/throat: 1mm (40 thou') 30° top cut, 1mm (40 thou') 45° seat, 1mm (40 thou') 60° bottom cut blended into throat.

1275 A series inlet boss modifications. (Remove shaded areas).

Road inlet port.

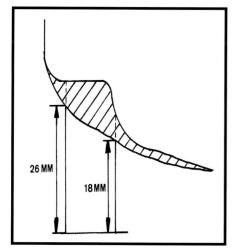

26 MM

18 MM

1275 A-series exhaust boss modification.
(Remove shaded areas).

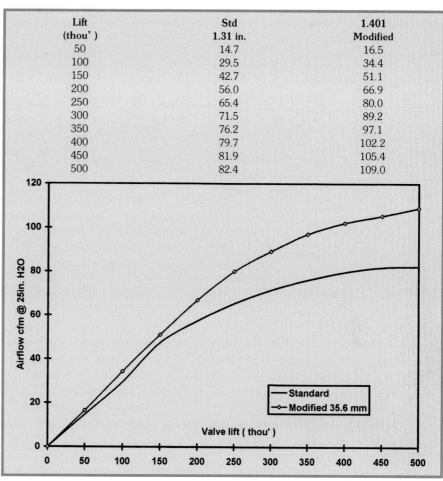

Lift (thou')	Std 1.31 in.	1.401 Modified
50	14.7	16.5
100	29.5	34.4
150	42.7	51.1
200	56.0	66.9
250	65.4	80.0
300	71.5	89.2
350	76.2	97.1
400	79.7	102.2
450	81.9	105.4
500	82.4	109.0

Inlet port flow @ 25 in. H2O. (1275 A-series).

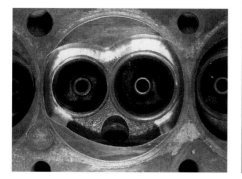

Race combustion chamber shape.

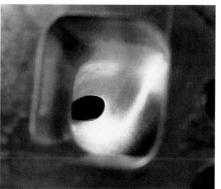

Road/race exhaust port.

FORD 351 CLEVELAND V8 (2V)

See Typical Flowbench Test Programme in Appendix.

ROVER V8 3.5 (SD1)

In standard form the valve throats are far too narrow. Recutting the 3-angle seats to the full valve diameters and opening and blending the throats to match, in conjunction with modifying the backs of the valves, adds about 35bhp at the wheels.

Fully modifying, by removing the guide bosses and opening the ports, adds another 10bhp.

The head really needs bigger valves to work effectively. However,

Lift (thou')	Std.	Throat and seat mod's	Full race 45.8 mm
50	18.2	20.4	22.8
100	37.3	42.1	46.0
150	56.0	63.5	70.5
200	72.9	83.6	93.4
250	88.0	100.5	112.7
300	97.1	111.4	128.0
350	104.8	119.0	139.4
400	111.1	123.6	147.6
450	112.1	126.7	153.7
500	113.2	128.5	157.0
550	---	---	161.0
600	---	---	162.4

Inlet port flow @ 25 in. H2O. (Rover V8).

Standard inlet port (top); modified port (above).

the inlet ports are too small to feed them and you find waterways and fresh air if you open them up too much! For road use we would recommend the opened throat heads as specified above. If you want more power change the cam.

Inlet seat/throat: 2mm (80

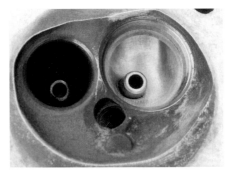

Race 45.8mm inlet valve throat and seat.

Race inlet port.

thou') 30° top cut, 1.3mm (52 thou') 45° seat, 1.5mm (60 thou') 60° bottom cut blended into throat.

Race: 1mm (40 thou') 30° top cut, 1.5mm (60 thou') 45° seat, 2mm (80 thou') 60° bottom cut blended into throat.

ROVER K-SERIES 16-VALVE 1.8 MGF

We've been undertaking a test/ evaluation/development programme on this 16-valve head with some interesting results.

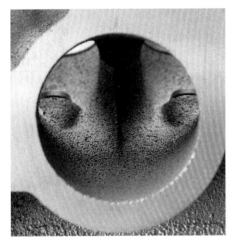

16-valve head standard inlet port before ...

... and after being modified too far and finding the waterways (visible below guides).

Lift (thou')	A	B	C	D	E	F	G
50	13.21	13.36	26.88	26.71	30.86	30.56	31.77
100	28.39	28.60	52.66	56.82	63.71	64.17	66.45
150	42.34	42.90	77.08	84.14	94.73	96.0	99.79
200	52.0	53.0	97.07	107.60	118.84	122.42	125.58
250	59.01	60.60	111.17	122.56	135.51	139.56	142.40
300	63.60	63.42	116.10	131.40	145.62	150.82	151.97
350	65.93	64.74	119.32	137.10	148.90	156.35	158.29
400	67.41	65.82	121.63	138.90	149.75	162.84	161.33
450	68.10	66.35	121.44	139.69	147.11	166.80	164.09

A= standard head 29mm inlet, left hand valve only
B= standard head 29mm inlet, right hand valve only
C = standard head 29mm inlets, both valves
"A", "B" and "C" conducted out of curiosity - do both ports flow the same and does the sum equal the total flow? The former, not far off. The latter, flow interference?
D = standard head 29mm inlets, 3-angle seats and throat modifications
E = Standard port, 31mm inlets, 3-angle seats and throats modified to suit
F = Penultimate port (see photo), fully modified ports and throats, 3-angle seats and 31mm tulip-shape valves
G = As "F" but with 31mm flatter-backed valves

Inlet port flow @ 25in H2O and corrected to 15 degrees C and 1013.25mb.

Valve shapes used during testing of 16-valve head (see table).

Comments -

'One step beyond' illustrated by photo showing breakthrough to the waterways.

Interesting to note the changes in flow due to the valve shapes (see photo), the flatter profiles favouring low to mid lift flows, while the tulip-shape starts working better at higher lifts.

Testing and experimentation continues ...

GENERAL

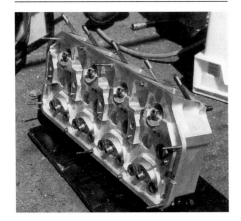

Hewn from Billet, these works of art help top fuel dragsters make such awesome horsepower.

Fastened to a blown 500cu.in. Hemi fed on nitromethane and unleashing 5000 to 5500 horsepower!

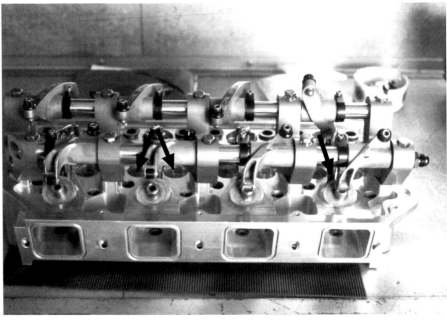

Huge inlet ports roller rockers and port injection (arrow) with dual plugs per cylinder (left).

The insert came adrift on these modified V8 heads, breaking a valve. Welding up the damage, fitting new inserts and a thorough re-work meant they were ready to race again.

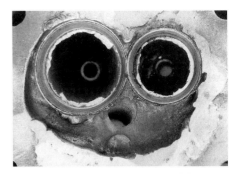

Finding third instead of fifth when on the rev limiter, helped a valve head decide to call it a day, wiping out the seats and damaging the chamber. Skilled alloy welding repaired and reclaimed the damage. New inserts were fitted, the seats recut and the chamber reworked to original shape (as right) and was then ready to race again ...

Appendix

COMPRESSION RATIO CALCULATIONS

The compression ratio (CR) of an engine is given by the following formula -

$$CR = \frac{\text{Swept volume} + \text{Clearance volume}}{\text{Clearance volume}}$$

Where -

Swept volume (Vs) is the capacity of one cylinder in cm³.

Clearance volume (Vc) comprises the cm³ above the top piston ring, any dish or valve cutouts in the piston, the gap between the top of the piston at TDC and the top of the block, the compressed head gasket volume and the combustion chamber volume.

For example -

1798cm³ block, 6.5cm³ dished

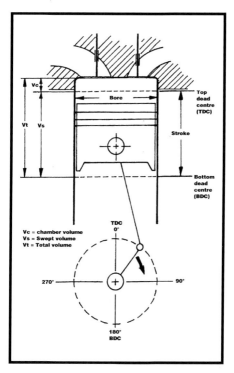

Volume measurements.

piston, 2cm³ above the piston, 4.5cm³ gasket volume and 43cm3 combustion chamber volume in the head.

Swept volume of one cylinder =
$$\frac{1798}{4} = 449.5cc$$

Clearance volume (Vc) =
$$6.5 + 2 + 4.5 + 43 = 56cm^3$$

$$CR = \frac{449.5 + 56}{56}$$

$$= 9.03:1 \ (\ 9.03 \text{ to } 1\ )$$

If the CR needs to be raised the new clearance volume required can be calculated as follows -

$$CR - 1 = \frac{\text{Swept volume}}{\text{Clearance volume}}$$

Therefore -
$$\text{Clearance volume} = \frac{\text{Swept volume}}{CR - 1}$$

For example -
The new CR. required is 10.5:1.

What chamber volume is needed to achieve this?

Using the same cylinder volume as previously (449.5cm^3) -

$$\text{Clearance volume} = \frac{449.5}{10.5 - 1}$$

$$\frac{449.5}{9.5}$$

$$= 47.32$$

As we know the clearance volume values as a 6.5cm^3 dish in the piston, 2cm^3 above the piston and 4.5cm^3 gasket volume, the chamber volume can be established by subtracting these values from the calculated clearance volume. For example -

$$\text{New chamber volume} = 47.32 - 6.5 - 2 - 4.5$$

$$= 34.32\text{cm}^3$$

HEAD SKIMMING (PLANING) CALCULATIONS

$$\text{New chamber depth required} = \text{original depth} \times \frac{\text{cm}^3 \text{ required}}{\text{original cm}^3}$$

For example -

cm^3 required is 34.32, original is 43cm^3 and the original chamber depth is 10mm.

$$\text{New chamber depth} = 10 \times \frac{34.32}{43}$$

$$= 7.98\text{mm}$$

Therefore skim 10 - 7.8 2.02mm (79 thou') from the head face.

GAS SPEED THROUGH INLET VALVES - EFFECTS OF ENGINE SIZE & CAMSHAFT SELECTION

$$\text{Peak power rpm} = \frac{\text{Gas speed} \times 5900 \times \text{valve area}}{\text{Cylinder volume}}$$

Where -

Gas speed is in feet/second
Valve area is in in^2
Cylinder volume is in cm^3

For example, a standard large inlet valve 1798cc MGB, peak power is at 5000rpm, valve area is 2.074in^2 and cylinder volume is 449.5cm^3.

$$\text{Gas speed} = \frac{\text{rpm} \times \text{cylinder volume}}{5900 \times \text{valve area}}$$

$$= \frac{5000 \times 449.5}{5900 \times 2.074}$$

$$= 183.7\text{ft/sec.}$$

With a standard cam and a 1950cc engine (487.5cm^3 per cylinder) the engine will peak at a lower rpm as the gas speed will be reached earlier -

$$\text{Peak rpm} = \frac{183.7 \times 5900 \times 2.074}{487.5}$$

$$= 4611\text{rpm.}$$

This is very close to the 4700rpm peak found on the rolling road.

If the engine is fitted with a tuned cam that increases maximum gas speed to 195ft/sec., then -

$$\text{Peak rpm} = \frac{195 \times 5900 \times 2.074}{487.5}$$

$$= 4895\text{rpm.}$$

Which, again, is close to the 5000rpm peak found on the rolling road.

TORQUE & HORSEPOWER CALCULATIONS

The relationship between torque and horsepower -

$$\text{Bhp} = \frac{\text{Rpm} \times \text{Torque}}{5252}$$

$$\text{Torque} = \frac{\text{Bhp} \times 5252}{\text{Rpm}}$$

Torque figures are measured directly from the engine by a dynamometer, and then mathematically converted to horsepower (bhp).

For example, if an engine produces 79bhp at 3000rpm, 107bhp at 4000rpm and 128bhp at 5200rpm. How does the torque output vary?

$$\text{Torque} = \frac{79 \times 5252}{3000} = 138.3\text{lb.ft}$$

$$\text{Torque} = \frac{107 \times 5252}{4000} = 140.5\text{lb.ft}$$

$$\text{Torque} = \frac{128 \times 5252}{5000} = 129.3\text{lb.ft}$$

From this you can see that the torque output curve is fairly flat - a good torquey (lots of grunt) engine. These figures are from a 1900 MGB with a fairly fast road cam. At 5252rpm bhp and torque must be equal. Torque produces acceleration, whereas bhp provides top speed (chant the mantra - horsepower is miles per hour!). Ideally, the rpm difference between peak torque and peak horsepower should be as large as possible (then the engine will pull like a steam train (and still rev well!), providing maximum flexibility.

TORQUE & ACCELERATION CALCULATIONS

If the total weight of the car and the engine's maximum torque are known then the 0 to 60mph acceleration can be estimated using the following formula -

$$0 \text{ to } 60\text{mph (seconds)} = \frac{2 \times W ^\wedge 0.6}{T}$$

Where -

W = weight of car

T = maximum torque in lb.ft

Don't forget to include the driver's weight, and the weight of fuel in the tank (approx. 7.5lb per Imperial gallon or 6.25lb per US gallon)

For example -

A car with fuel weighs 2395.8lb, plus driver (say, 150lb) and maximum torque is 100lb.ft.

$$0 \text{ to } 60\text{mph} = \frac{2 \times 2545.8 ^\wedge 0.}{100}$$

= 10.6 seconds.

Increasing torque will reduce the 0 to 60mph time. For instance, the engine used as the example for the torque and horsepower maths produced a maximum torque of 145lb.ft.

$$0 \text{ to } 60\text{mph} = \frac{2 \times 2545.8 ^\wedge 0.6}{145}$$

= 8.46 seconds.

Suppose the car weight is reduced by, say, 200lb, what would be the effect on the 0 to 60mph acceleration time?

$$0 \text{ to } 60\text{mph} = \frac{2 \times 2345.8 ^\wedge 0.6}{145}$$

= 8.05 seconds.

If the maximum bhp is known, "a rule of thumb" method of establishing the increase in top speed is as follows:

$$\text{Max. speed} = \sqrt[3]{\frac{new\ bhp}{old\ bhp}} \times old\ maximum\ speed$$

For example -

A car has 85bhp at the wheels and a top speed of 108mph, but what would be the new maximum speed with, say, 110bhp at the wheels?

$$\text{Max. speed} = \sqrt[3]{\frac{110}{85}} \times 108$$

$$= \sqrt[3]{1.294} \times 108$$

$$= 1.089 \times 108$$

$$= 117.7\text{mph}$$

An alternative example -

How much bhp at the wheels is necessary to do 145mph?

$$\text{New bhp at wheels} = \frac{new\ mph}{old\ mph}^3 \times old\ bhp$$

$$= \frac{145}{108}^3 \times 85$$

$$= 2.42 \times 85$$

$$= 205\text{bhp at the wheels.}$$

FUEL SUPPLY CALCULATION

The following formula can be used to arrive at an approximate idea of the engine's fuel supply needs, in order to be able to size the fuel pump correctly.

Fuel consumption (lb per hour) = maximum engine horsepower x B.S.F.C

B.S.F.C stands for Brake Specific Fuel Consumption, which is given in pounds of fuel used per horsepower per hour. Most engines run figures between 0.45 (very efficient) and 0.6 (rather thirsty) so 0.5 is a reasonable average value for a roadgoing vehicle.

One gallon of fuel (UK Imperial measure) weighs around 7.5lb.

Gallons per hour =
(Max. engine hp x 0.5) / 7.5

Here's an example for a fairly powerful car - (225bhp x 0.5) / 7.5 = 15 gallons per hour. In this example you'll need to select a pump that can supply above 15gph - typically 18 gallons per hour.

For conversion to US gallons multiply the result by 1.20095 - in this example it would be 15 x 1.20095 = 18 gallons per hour.

DECIPHERING CAMSHAFT SPECIFICATIONS

Most manufacturers supply the camshaft specification as a table of numbers, from which the timing figures are the ones we can work with.

For example, cam timing of 16/56 51/21 means the inlet valve starts to open at 16 degrees Before Top Dead Centre (BTDC) and closes 56 degrees After Bottom Dead Centre (ABDC). The exhaust valve starts to open 51 degrees before BDC and closes 21 degrees After Top Dead Centre (ATDC).

Overlap

The overlap period where both inlet and exhaust are open together is calculated by adding the first (inlet valve opens) and the last (exhaust valve closes) figures together. For example -

16/56 51/21

16 + 21 = 37 degrees

So this cam has 37 degrees overlap.

Duration

The duration (the time the valves are open) is calculated as follows -

16 + 56 + 180 = 252 degrees duration for the inlet

51 + 21 + 180 = 252 degrees duration for the exhaust

(180 degrees is the one half of a stroke that each induction or exhaust cycle takes).

Lobe centre angle

The lobe centre angle (point of peak cam lift) for the cam is calculated as follows -

$$\frac{Duration}{2} - \text{valve opening point for the inlet or closing point for the exhaust}$$

$$= \frac{252 - 16}{2}$$

= 110 degrees for the inlet

Or -

$$= \frac{252 - 21}{2}$$

= 105 degrees for the exhaust

The inlet full lift point provides a starting point for cam installation if none is specified by the manufacturer. You may find these figures may

vary from the camshaft manufacturer's recommended installation figures, ie. the inlet peak cam lift may work out (mathematically) at 110 degrees (ATDC) but the recommended installation is 108 degrees, two degrees retarded (the inlet valve reaches peak lift sooner). This discrepancy is usually because, during testing, the cam was found to work best timed at the 108 degree setting.

The foregoing calculations are only true for symmetrical cams with the same opening and closing rates on the lobe. With asymmetric cams, peak cam lift can occur either side of this figure, depending on the lobe profile. In such a case it's best to follow the manufacturer's recommendations.

Lobe separation angle

The lobe separation angle (LSA) is calculated as follows -

$$\frac{\text{inlet lobe centre angle} + \text{exhaust lobe centre angle}}{2}$$

$$= \frac{105 + 110}{2}$$

= 107.5 degrees

Of course, all these figures are only useful as guidelines due to the differences in the basis of advertised timing figures used by the various manufacturers. It all depends upon what valve lift, if any, they choose to base their figures upon.

CAM TIMING

The correct installation and timing of the camshaft is, undoubtedly, one of the most vital steps toward achieving a powerful, smooth and progressive engine. You may have been led to believe that it is simply a matter of

fitting the camshaft timing gear and aligning the dots, as it says in the manual. Wrong! In the majority of cases the tolerances of the various components in the drive train can add up to the camshaft being a considerable way off the manufacturer's intended setting. This particularly applies to reground aftermarket profiles where, in order to reprofile a worn out original, the cam timing can become changed due to the nature of the reprofiling work. Also, the correct installation position can be very different from standard if an aftermarket high performance cam is being fitted, so note that the recommended installation position is usually given on the specification sheet that comes with the cam from the manufacturer.

As the camshaft is really the 'brain' of the engine, in that it controls the timing and rate of the opening and closing of the valves during the four stroke cycle, any significant variation from the intended timing can considerably alter the engine's power delivery characteristics. Small variations in timing may have no affect on engine performance, but you never can tell. So, correct timing of the cam is **essential** to avoid disappointment in the engine's performance.

There are various methods for timing a cam and the most common for a cam-in-block engine is by using the inlet full lift position, which we will now describe (though this assumes the engine is not still in the vehicle). This technique can also be used for cam timing on most overhead cam engines, though you may need to rig up a custom extension to allow the dial gauge to reach the top of the valve cap/follower/bucket, so valve lift can be measured.

Correctly timing the camshaft requires a few pieces of equipment.

These comprise a dial gauge (on a magnetic stand for preference), a large 360 degree protractor (called the timing disc or degree wheel - available from cam companies) and a short length of wire that can be bent to form a pointer.

With the camshaft installed, fit the cam sprocket/pulley and timing chain - there's no need to fit the camshaft sprocket/pulley retaining nut and washer yet - and align the dots (or whatever) on the crankshaft and camshaft sprockets, as given in the workshop manual relevant to the engine.

The timing disc can then be fitted to the front of the crank using the bolt for the front pulley (you may need to use a couple of thick washers as packing behind the disc, otherwise the bolt will 'bottom' before it clamps the disc). Very lightly nip up the bolt with your fingers (the disc will need to be rotated in order to zero it later). The wire pointer should then be attached to the block close to the timing disc (wrapping it securely around a protruding water pump bolt works well) and bent to protrude over the edge of the disc so that it's aligned with the graduations.

At this point life can be made a little easier if two of the flywheel bolts are partly threaded into the back of the crank (opposite each other). These can then provide assistance in turning the engine back and forth, by putting a long screwdriver or bar between them to apply leverage. Not that your beautifully assembled short motor should be difficult to turn over, it's simply that it can get a bit stiff occasionally due to the engineering phenomena called "stick-slip," and Sod's Law means it usually occurs when a very small movement is called for!

We now need to find, accurately,

Finding true TDC of No.1 piston.

Top Dead Centre for number one piston. With the piston visually at top dead centre in its bore, position the dial gauge stand on the top of the block and adjust it so the gauge is touching in the middle of the piston (if the head's still on you can go down through a plug hole to the piston crown). Centreing is necessary to limit any errors in the readings caused by the piston rocking in the bore. Piston TDC can now be established by turning the engine back and forth several times and zeroing the dial gauge when the piston is at its highest point in the bore.

The problem is that this point may not be the exact piston TDC position. You may have noticed that while continuing to turn the crank, the reading on the dial gauge remains at zero through a few degrees of crank revolution before the piston moves down the bore again and the reading changes. This is due to the piston dwell period at TDC, which is a function of crank stroke and con rod

length, and varies from engine to engine.

You can generally get a feeling for how long this dwell period is when turning the engine to and fro and you should set the degree wheel and pointer to zero at its mid point. However, if you wish to be deadly accurate (and you should), there's a bit more to do yet !

Rotate the engine backwards through about one quarter of a revolution, then forwards until the dial gauge reads 0.001in/0.025mm before the zero position. Make a note of the reading on the timing disc (say, 3 degrees before TDC). Continue turning the engine forwards until the dial gauge moves off zero and reads 0.001 inch/0.025 mm again (say, 1 degree after TDC). Make a note of this reading. Add the two readings together and divide by two (3 + 1 = 4 / 2 = 2). Carefully move the disc (backwards in this example) until the pointer is positioned at 2 degrees after TDC, and tighten the nut holding the timing disc. Then turn the engine backwards one quarter of a turn, then forwards in order to re-check the timing disc is positioned properly with the pointer showing TDC. If you had to slacken it off again, tighten up the nut to clamp the disc securely, being careful not to move or dislodge anything. Double-check the readings, and then this part is completed.

A lightly oiled cam follower should be fitted into position in the block on the intake lobe of the number one cylinder. Slot in a pushrod from the top of the block and give it a wiggle to check it is seated in the follower correctly. The dial gauge and stand needs to be set up so that it seats in the little cup at the top of the pushrod, so the pushrod and the whole assembly are positioned as they would be with the engine fully

Using a dial gauge to find full lift on inlet lobe of No.1 cylinder.

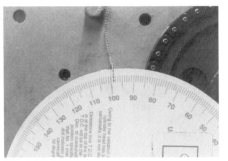

Half a thou' before full lift, protractor shows 99° ATDC ...

... half a thou' after full lift, protractor reading 110°. Cam is positioned at 105.5° ...

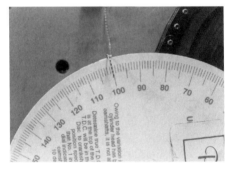

... while what was wanted was full lift at 100°. With a 5° key fitted cam was correctly timed at required 100° (half a thou' before = 95°, half a thou' after = 105°).

A selection of offset keys for the camshaft sprocket.

Offset key to advance camshaft by 5°.

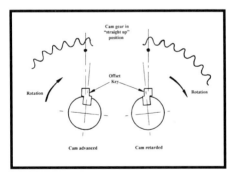

The effect of offset keys.

assembled.

From now on, always take any readings while turning the engine forwards in the direction of rotation. If you miss one, or make a mistake, always turn the engine backwards at least a quarter to half a turn before going forwards again, or the slack in the timing chain will cause no end of problems with the readings.

Slowly turn the engine forwards until the camshaft reaches maximum lift and then zero the dial gauge. Now turn the engine backwards about a quarter of a turn, then slowly forwards again until the dial gauge reads 0.0005in/0.01mm) before full lift; it needs to be half a thou' spot on. Make a note of the number of degrees after TDC shown on the degree wheel. Slowly continue turning the engine forwards, past full lift, until the dial gauge reads 0.0005in/0.01mm after full lift, spot on again. Make a note of the new reading on the degree wheel. Adding these two readings together,

and then dividing by two, gives the number of degrees after TDC that full lift of the camshaft occurs at as currently installed. This can then be compared with the figure given on the specification sheet for that particular cam.

For example, for those engines where the cam should be timed in at

106 degrees after TDC, if the cam timing is right the average of the two readings taken should be 106. If it is within a degree or so either way then you are very fortunate in that nothing more needs to be done, the cam is correctly positioned.

If, as is more often the case, the resulting full lift position is not what is required, steps must be taken to remedy the situation. If the number you have calculated is low, say, 104

Duplex adjustable cam sprocket. Slacken the Allen head fixings to make easy camshaft timing changes.

Caution! - Always re-check the timing again after making any adjustments to ensure that you didn't put the key in the wrong way around by mistake.

By far the easiest means of altering the cam timing is by using an adjustable or vernier sprocket/pulley. With the adjustable sprocket/pulley the locking bolts are slackened allowing the cam to be turned without moving the crank, so you simply have to set the crank to the correct number of degrees that full cam lift should be, and then turn the cam until the dial gauge shows full cam lift. Lock up the retainers, and the job's done, bar a final check to ensure the timing is definitely correct. With the vernier, it is a matter of putting the pin in another hole and re-checking the timing until correct.

Crankshaft sprockets are available with several preset offset slots for the key already machined in, for example in the straight up and three degrees advanced or retarded positions. Again, a little juggling of the permutations is all that is required to arrive at the right result.

Vernier (pin and holes) or adjustable (Allen head bolts and slots) sprockets do make life easier, but cost considerably more than a set of offset keys. So, "you pays your money and you takes your choice."

In practice, cam timing is far easier than this long-winded description implies. The accompanying photos should make the procedure a little clearer. It probably seems a complex and daunting task to even attempt but, given a little patience and thought, degreeing a cam is really quite straightforward and the end result is very rewarding.

degrees, then the cam is advanced. If the number is high, say, 110 degrees, then it is retarded. To advance the cam it will have to be turned slightly forwards, in the direction it rotates and vice versa to retard it.

When using the standard cam sprocket/pulley the timing can be changed by using an offset key in the cam keyway in place of the standard one. These keys are available with offset increments from one to nine crank degrees (0.5 to 4.5 cam degrees - remember, one cam revolution takes two crank revolutions). Also, with moving the chain around by a single tooth on the cam sprocket giving an eighteen crank degree change, reversing the key allows increments back from the eighteen. The permutations are all there, they just require a little juggling to get the correct result.

TYPICAL FLOW BENCH TEST PROGRAMME

FLOWBENCH ANALYSIS

DATE: 17-20 January 1995   **TEST:** Ford 351 Cleveland V8 2V heads   **TESTED:** DKG/PB.   **INLET:** 2.041in.

AIRFLOW IN CUBIC FEET PER MINUTE

Flow is at 25in water and corrected to 15°C and 1013.25mb

LIFT (thou)	A	B	C	D	Change (A to D)
50	22.26	24.37	24.83	25.78	+ 15.6%
100	48.93	55.16	57.00	58.65	+17.6%
150	73.2	82.52	84.39	87.42	+19.4%
200	99.54	113.15	118.59	120.84	+21.4%
250	124.69	140.52	143.72	147.00	+17.9%
300	145.69	161.29	166.68	170.91	+17.3%
350	156.46	171.42	177.73	182.23	+16.5%
400	163.10	178.27	185.21*	187.70	+15.1%
450	172.77	183.23	183.18	188.61*	+9.2%
500	180.29	183.28*	180.58	187.22	+3.3%
550	182.83*	178.59	178.73	186.51	+2.5%
600	182.06	176.67	176.44	185.40	+1.8%
650	182.06	176.16	174.38	184.20	+1.2%

* Indicates max. value.
A = Standard head.
B = 3 angle seat (1.3 mm) and modified valve shape. Standard port.
C = Short turn radius improved. Throat smoothed and blended.
D = Fully modified port with guide boss removed.

Comments

The rectangular port is too small to achieve a good blend into throat area (even with hump on port floor) with this valve size. As the valve flow is improved the flow finds it increasingly difficult to negotiate the short turn, evidenced by the flow dropping off with increasing lift. Thinness of port walls precludes any further material removal.

The fact that flow does not increase significantly from the original maximum - merely being reached at lower lift values with the additional modifications - indicates that the head is port restricted with this size valve.

Options

Use same type heads but with smaller diameter valve (1.9in?) to improve port to throat blend (good for torque).

Investigate big port 4V heads for high rpm power production - but significant loss of low end torque with current engine specification highly probable.

Alternatively, investigate aluminium head options offered by Ford Motorsport (USA).

FLOWBENCH ANALYSIS

DATE: 17-20 January 1995   **TEST**: Ford 351 Cleveland V8 2V heads   **TESTED**: DKG/PB.   **EXHAUST**: 1.655in

AIRFLOW IN CUBIC FEET PER MINUTE

Flow is at 25in water and corrected to 15°C and 1013.25mb

LIFT (thou)	A	B	Change
50	21.26	23.43	+10.2%
100	37.8	44.36	+17.2%
150	61.72	70.63	+14.4%
200	81.46	91.29	+12.0%
250	102.69	111.81	+8.9%
300	117.20	126.55	+7.9%
350	127.54	134.49	+5.4%
400	134.70	140.28	+4.1%
450	139.31	144.67	+3.8%
500	142.18	148.02	+4.1%
550	143.26	149.53	+4.4%
600	144.90*	151.00*	+4.2%

* Indicates max. value.
A = Standard head.
B = 3 angle seat (1.3 mm) and modified valve shape. Guide boss reduced, throat smoothed and blended, modified port.

ALSO FROM VELOCE PUBLISHING -

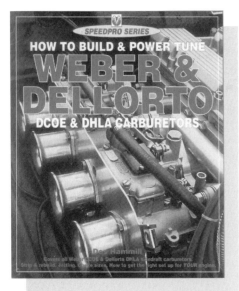

HOW TO BUILD & POWER TUNE WEBER & DELLORTO DCOE & DHLA CARBURETORS
- 2ND Edition
by Des Hammill

ISBN 1 901295 64 8
Price £14.99*

A book in the *SpeedPro* series. All you could want to know about the world's most famous and popular high-performance sidedraught carburetors. Strip & rebuild. Tuning. Jetting. Choke sizes. Application formula gives the right set-up for *your* car. Covers all Weber DCOE & Dellorto DHLA carburetors.

CONTENTS
COMPONENT IDENTIFICATION: The anatomy of DCOE & DHLA carburetors • DISMANTLING: Step-by-step advice on dismantling. Assessing component serviceability • DIFFICULT PROCEDURES: Expert advice on overcoming common problems in mechanical procedure • ASSEMBLY: Step-by-step advice on assembly. Fuel filters. Ram tubes. Fuel pressure • SETTING UP: Choosing the right jets and chokes to get the best performance from *your* engine • FITTING CARBURETORS & SYNCHRONISATION: Covers alignment with manifold and balancing airflow • FINAL TESTING & ADJUSTMENTS: Dyno and road testing. Solving low rpm problems. Solving high rpm problems. Re-tuning.

THE AUTHOR
Des Hammill has a background in precision engineering and considers his ability to work very accurately a prime asset. Des has vast experience of building racing engines on a professional basis and really does know how to get the most out of a Weber or Dellorto carburetor. Having lived and worked in many countries around the world, Des currently splits his time between the UK and New Zealand.

SPECIFICATION
Softback • 250 x 207mm (portrait) • 112 pages • Over 140 black & white photographs and line illustrations.

RETAIL SALES
Veloce books are stocked by or can be ordered from bookshops and specialist mail order companies. Alternatively, Veloce can supply direct (credit cards accepted).

* Price subject to change.

Veloce Publishing Plc, 33 Trinity Street, Dorchester, Dorset DT1 1TT, England. Tel: 01305 260068/ Fax: 01305 268864/ E-mail: veloce@veloce.co.uk

Visit Veloce on the Web - www.veloce.co.uk

ALSO FROM VELOCE PUBLISHING -

HOW TO BUILD & POWER TUNE DISTRIBUTOR-TYPE IGNITION SYSTEMS
by Des Hammill

ISBN 1 874105 76 6
Price £9.99*

A book in the **SpeedPro** series. Expert practical advice from an experienced race engine builder on how to build an ignition system that delivers maximum power reliably. A lot of rubbish is talked about ignition systems and there's a bewildering choice of expensive aftermarket parts which all claim to deliver more power. Des Hammill cuts through the myth and hyperbole and tells readers what *really* works, so that they can build an excellent system without wasting money on parts and systems that simply don't deliver.

Ignition timing and advance curves for modified engines is another minefield for the inexperienced, but Des uses his expert knowledge to tell readers how to optimise the ignition timing of *any* high-performance engine.

The book applies to all four-stroke gasoline/petrol engines with distributor-type ignition systems, including those using electronic ignition modules: it does not cover engines controlled by ECUs (electronic control units).

CONTENTS
Why modified engines need more idle speed advance • Static idle speed advance setting • Estimating total advance settings • Vacuum advance • Ignition timing marks • Distributor basics • Altering rate of advance • Setting total advance • Quality of spark •

THE AUTHOR
Des Hammill has a background in precision engineering and considers his ability to work very accurately a prime asset. Des has vast experience of building racing engines on a professional basis. Having lived and worked in many countries around the world, he currently splits his time between the UK and New Zealand.

SPECIFICATION
Softback • 250 x 207mm (portrait) • 64 pages • Over 70 black & white photographs and line illustrations.

RETAIL SALES
Veloce books are stocked by or can be ordered from bookshops and specialist mail order companies. Alternatively, Veloce can supply direct (credit cards accepted).

* Price subject to change.

Veloce Publishing Plc, 33 Trinity Street, Dorchester, Dorset DT1 1TT, England. Tel: 01305 260068/ Fax: 01305 268864/E-mail: veloce@veloce.co.uk

ALSO FROM VELOCE PUBLISHING -

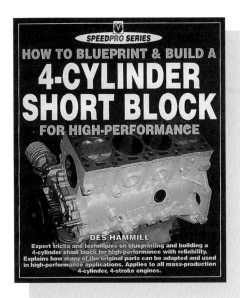

A book in the **SpeedPro** series.

• Applies to all 4-cylinder car engines (except diesel & two-stroke).
• Essential reading for millions of car owners looking for more power.
• Expert advice in non-technical English accompanied by clear photos & line illustrations.
• Saves money by eliminating techniques that don't work and by maximising the use of standard components.
• Written by a professional competition engine builder.

HOW TO BLUEPRINT & BUILD A 4-CYLINDER SHORT BLOCK FOR HIGH PERFORMANCE
by DES HAMMILL

ISBN 1 874105 85 5
Price £13.99 *

CONTENTS
A complete practical guide on how to blueprint (optimize all aspects of specification) any 4-cylinder, four-stroke engine's short block to obtain maximum performance and reliability without wasting money on over-specced parts. Includes choosing components, crankshaft & conrod bearings, cylinder block, connecting rods, pistons, piston to valve clearances, camshaft, engine balancing, timing gear, lubrication system, professional check-build procedures and much more. Index.

SPECIFICATION
Paperback. 250 X 207mm (portrait). 112 pages. Around 200 black & white photographs/illustrations.

RETAIL SALES
Veloce books are stocked by or can be ordered from bookshops and specialist mail order companies. Alternatively, Veloce can supply direct (credit cards accepted).

* Price subject to change.

Veloce Publishing Plc, 33 Trinity Street, Dorchester, Dorset DT1 1TT, England. Tel: 01305 260068/ Fax: 01305 268864/E-mail: veloce@veloce.co.uk

ALSO FROM VELOCE PUBLISHING -

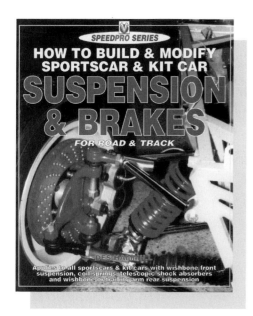

HOW TO BUILD & MODIFY SPORTSCAR & KIT CAR SUSPENSION & BRAKES
by Des Hammill

ISBN 1 901295 08 7
Price £14.99*

A book in the **SpeedPro** series.

• Applies to all two seater sportscars and kit cars with wishbone front suspension, coil springs, telescopic shock absorbers and wishbone or trailing arm rear suspension.
• Basic information applies to all cars.
• Written in a clear understandable style with over 100 detailed original diagrams.
• Cuts through the mystique and confusion surrounding suspension and handling improvements.
• Ideal for the home mechanic.

• Applies to road and track applications.
• Des Hammill is an engineer and a professional race car builder with many years of practical experience.

CONTENTS
Chassis integrity • Suspension geometry • Ride height • Negative camber, castor and Kingpin inclination • Springs and shock absorbers • Brakes • Setting up the car • Testing, alterations and adjustments • Index

SPECIFICATION
Paperback. 250 x 207mm (portrait). 112 pages. Over 100 detailed original diagrams.

RETAIL SALES
Veloce books are stocked by or can be ordered from bookshops and specialist mail order companies. In case of difficulty we can supply direct (credit cards accepted).

* Price subject to change.

Veloce Publishing Plc, 33 Trinity Street, Dorchester, Dorset DT1 1TT, England. Tel: 01305 260068/ Fax: 01305 268864/E-mail: veloce@veloce.co.uk

Visit Veloce on the Web - www.veloce.co.uk

SPEEDPRO SERIES

ALSO FROM VELOCE PUBLISHING -

HOW TO CHOOSE CAMSHAFTS & TIME THEM FOR MAXIMUM POWER
by Des Hammill

ISBN 1 901295 19 2
Price £10.99*

A book in the **SpeedPro** series. Explains in simple language how to choose the right camshaft/s for *YOUR* application and how to find the camshaft timing which gives maximum performance.
• Also explained are all aspects of camshaft design and the importance of lobe phasing, duration & lift.
• Applies to all 4-stroke car-type engines with 4, 5, 6 or 8 cylinders.
• Des Hammill is an engineer and a professional race engine builder with many years of experience.
• Avoids wasting money on modifications that don't work.
• Applies to road and track applications.

CONTENTS
Introduction • Using This Book &
Essential Information • Chapter 1: Terminology • Chapter 2: Choosing the Right Amount of Duration • Chapter 3: Checking Camshafts • Chapter 4: Camshaft Timing Principles • Chapter 5: Camshaft Problems • Chapter 6: Timing Procedure - Cam-in-Block Engines • Chapter 7: Camshaft Timing Procedure - S.O.H.C. Engines • Chapter 8: Camshaft Timing Procedure - T.O.H.C. Engines • Chapter 9: Engine Testing • Index.

THE AUTHOR
Des Hammill has a background in precision engineering and places great emphasis on accuracy. Des has vast experience of building all types of engine for many categories of motor racing. Having lived in many countries around the world, Des and his wife, Alison, currently live in Devon, England.

SPECIFICATION
Softback • 250 x 207mm (portrait) • 64 pages • 150 black & white photographs & line illustrations.

RETAIL SALES
Veloce books are stocked by or can be ordered from bookshops and specialist mail order companies. Alternatively, Veloce can supply direct (credit cards accepted).

* Price subject to change.

Veloce Publishing Plc, 33 Trinity Street, Dorchester, Dorset DT1 1TT, England. Tel: 01305 260068/ Fax: 01305 268864/E-mail: veloce@veloce.co.uk

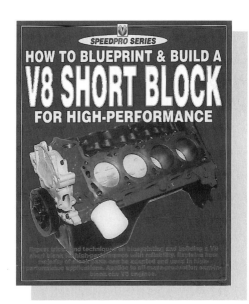

HOW TO BLUEPRINT & BUILD A V8 SHORT BLOCK FOR HIGH PERFORMANCE
by Des Hammill

ISBN 1 874105 85 5
Price £13.99*

A book in the **SpeedPro** series. Expert practical advice from an experienced race engine builder on how to build a V8 short engine block for high performance use using mainly stock parts - including crankshaft and rods. A short block built using Des' techniques will be able to deliver serious high performance with real reliability. Applies to all sizes and makes of V8 engine with overhead valves operated by pushrods.

CONTENTS
Selecting a suitable short block • Stripdown • Checking critical sizes • Choosing replacement (including non-stock) parts • Cleaning of block & parts • Checking condition of all parts • Crack testing • Remachining • Balancing • Camshaft & lifters • 'Check fitting' engine build technique • Bearing crush • Final rebuild • Checking true top dead centre • Additional degree markings for camshaft & ignition timing • ACcurate camshaft timing • oil pan requirements.

THE AUTHOR
Des Hammill has a background in precision engineering and places great emphasis on accuracy. Des has vast experience of building all types of engine for many categories of motor racing. Having lived in many countries around the world, Des and his wife, Alison, currently live in Devon, England.

SPECIFICATION
Softback • 250 x 207mm (portrait) • 112 pages • over 180 black & white photographs & line illustrations.

RETAIL SALES
Veloce books are stocked by or can be ordered from bookshops and specialist mail order companies. Alternatively, Veloce can supply direct (credit cards accepted).

* Price subject to change.

Veloce Publishing Plc, 33 Trinity Street, Dorchester, Dorset DT1 1TT, England. Tel: 01305 260068/ Fax: 01305 268864/E-mail: veloce@veloce.co.uk

Visit Veloce on the Web - www.veloce.co.uk

Index